Fundamentals of Energy Dispersive X-ray Analysis

Butterworths Monographs in Materials

The intention is to publish a series
of definitive monographs written
by internationally recognized
authorities in subjects at the
interface of the research interests
of the academic materials scientist
and the industrial materials
engineer.

Already published

Amorphous metallic alloys
Control and analysis in iron and
    steel making
Die casting metallurgy
Eutectic solidification processing
Introduction to the physical
    metallurgy of welding
Metals resources and energy
Powder metallurgy of super alloys

Forthcoming titles

Continuous casting of aluminium
Mechanical properties of ceramics
Metallurgy of high speed steels
Microorganisms and metal recovery
Residual stresses in metals

Butterworths Monographs in Materials

Fundamentals of Energy Dispersive X-ray Analysis

John C. Russ
Research Associate and Manager
Analytical Laboratories, Engineering Research Division,
North Carolina State University, Raleigh, NC, USA

Butterworths
London Boston Durban Singapore Sydney Toronto Wellington

First published, 1984

© Butterworths & Co (Publishers) Ltd., 1984

British Library Cataloguing in Publication Data

Russ, John C.
    Fundamentals of energy dispersive X-ray analysis.
    – (Butterworths monographs in materials)
    1. Solids – Spectra   2. X-ray spectroscopy
    I. Title
    530.4'1   QC176.8.X/

ISBN 0–408–11031–7 ✓

Library of Congress Cataloguing in Publication Data

Russ, John C.
    Fundamentals of energy dispersive X-ray analysis.

    (Butterworths monographs in materials)
    Bibliography: p.
    Includes index.
    1. X-ray spectroscopy.   I. Title.   II. Series.
    QD96. X2R87 1984   537.5'35   83-23216
    ISBN 0–408–11031–7

Typeset by First Page Ltd, Watford, Herts.
Printed and bound by Robert Hartnall Ltd, Bodmin,
Cornwall.

Contents

Chapter 1

X-Ray Emission

X-rays are photons of electromagnetic radiation, like light, radio waves, and at the other end of the spectrum γ-rays. They are distinguished by their wavelength range, generally about 0.1 to 100 angstroms, and their origin from events in the structure of atoms. Most of the time, if you tell someone you work with X-rays, they will imagine you as a medical doctor looking at broken bones or a dentist at decayed teeth. As Roentgen, who discovered and named them observed, X-rays penetrate much further through ordinary things (like people) than visible light does, and this means that images formed with X-rays can be used to reveal internal structure. Industrial uses of X-radiography also abound for inspection of internal flaws in critical parts.

The analytical uses of X-rays include diffraction, which permits the study of crystalline order of materials and even large organic molecules like DNA, and emission or fluorescence spectrometry, which we shall discuss here. This is the measurement (spectroscopy would simply be the observation) of the energy or wavelength, and the number of X-rays emitted when atoms are excited, from which we can obtain compositional information in terms of the atomic species present.

Figure 1.1 An excited atom.

What happens when atoms get excited? First we should recall the simple Bohr picture of the atom, with electron shells or orbits.

Characteristic X-rays (ie. ones that can be identified with the element that produced them) are emitted when an electron jumps from one shell to another, lower energy one. For this to be possible, of course, we first must have an empty spot and this process of electron removal, or ionization, is called excitation. It can be accomplished with electron beams, or other charged particles such as protons or alpha particles, or by photons such as other X-rays or γ-rays (which originate in nuclear, rather than atomic processes but may have similar energies as X-rays). These various forms of excitation have very different probabilities of creating electron vacancies (called the ionization cross section), and furthermore these probabilities vary in complex ways with energy, atomic number, and other conditions, in ways that will be considered when appropriate. But all that is needed is a source of energy greater than the binding energy of the electron in the shell, which ranges from a few hundred electron volts for the inner shell (the only shell) of light elements, to about a hundred keV. for the inner shell of high atomic number elements.

The shells are named, starting from the inside (most strongly bound): K, L, M, N, etc., and the X-rays emitted when a vacancy in a particular shell is refilled carry the same designation. Historically, when the K shell was being studied and the energies measured, there was some doubt that it was truly the innermost one, and so by starting with the letter K, room was preserved to add others if needed.

If a K electron is removed from its shell, and the vacancy is filled by an L electron dropping into the lower energy site, the excess energy is released in the form of an X-ray photon called the K-α. If the electron comes from the M shall, the X-ray is a K-β. The Greek letters and numbers used with them identify the shells and

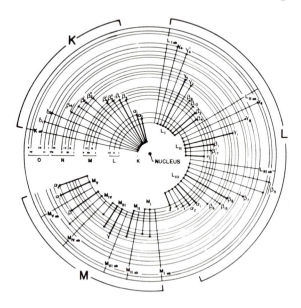

Figure 1.2 Schematic diagram of electron shells and the transitions giving rise to the principle X-ray emission lines.

subshells (because of course there are really three L, five M, and so forth) involved in the transition. Unfortunately when the early measurements were made and this nomenclature assigned, the detailed structures of the atoms were not well understood, and so the designations (α-1, β-3, γ-2 and so on) have more to do with the relative intensity of the line than with an orderly sequence of shells. Figure 1.2 shows the transitions involved in many of the more prominent lines.

The intensities are in turn controlled by the selection rules that govern the relative probabilities of the various transitions (Bambynek, 1972), and of course when outer shells are involved, the extent to which a particular shell has been filled with electrons. The quantum numbers for spin and angular momentum define the subshells, and the exclusion principle governs the number of electrons needed to fill each shell. The correspondence of the classical shell designations to the quantum mechanical notation in which n is the principal or energy quantum number, l the orbital angular momentum quantum number, j the spin quantum number, and m the magnetic quantum number is shown in the table below. The Pauli exclusion principle prohibits two electrons from having an identical set of quantum numbers, and thus gives rise to the total permitted number of electrons in each shell.

Table 1.1 Electron shell and quantum number designations

X-ray Shell	No. of electrons	Quantum n	numbers l	j	s-p-d notation
K	2	1	0	1/2	1s 1/2
L I	2	2	0	1/2	2s 1/2
L II	2	2	1	1/2	2p 1/2
L III	4	2	1	3/2	2p 3/2
M I	2	3	0	1/2	3s 1/2
M II	2	3	1	1/2	3p 1/2
M III	4	3	1	3/2	3p 3/2
M IV	4	3	2	3/2	3d 3/2
M V	6	3	2	5/2	3d 5/2
N I	2	4	0	1/2	4s 1/2
N II	2	4	1	1/2	4p 1/2
NIII	4	4	1	3/2	4p 3/2
N IV	4	4	2	3/2	4d 3/2
N V	6	4	2	5/2	4d 5/2
N VI	6	4	3	5/2	4f 5/2
N VII	8	4	3	7/2	4f 7/2
etc.					

We will consider the factors affecting the particular line intensities later. Generally, the ratios of intensity among the more important lines are fairly constant. For instance, the K-α-1 (transition from L III to K) is twice as intense (happens twice as often) as the K-α-2 (transition from L II to K —remember that the

L I - K transition is forbidden). The summation of the various K-β (M
-K transitions) intensities (probabilities) is about one fifth of the
total K-α (L - K transitions) intensities, for elements high enough
in atomic number that the M shell has been filled. The total intensity
in the L-α -1 and -2 lines (M V and M IV to L III transitions) is
about twice that of the L-β -1 (M IV to L II), but for the outer
shells with their many subshells, the line designations, yet alone
their intensities, become somewhat mixed up.

It is not usually necessary to remember these ratios, since most
tables used for peak identification (which may be built into the
software of modern computerised systems) will include the ratios, and
anyway they change from element to element as outer shells are filled,
and from sample to sample because the emitted intensities we are
discussing here are further modified by absorption in the matrix
before we can count them. Also, it is rarely important to consider all
of the possible emission lines. As few as 5 or 6 are usually adequate
for identification, and are used almost exclusively for quantitative
work. More, but not more than a total of about 30 (comprising all with
relative intensities above 1%), may be taken into account when
correcting for peak overlaps in which a minor line from a major
element has a wavelength (energy) close to that of a major line from a
trace element, and obscures it in the spectrum.

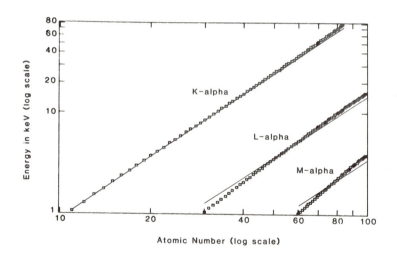

Figure 1.3 Logarithmic plot of X-ray energies (K- L- and M-α lines)
compared to Moseley's Law.

The foundation for X-ray analysis to determine composition was
laid by Moseley (1913), who observed that the frequency of X-rays from
different elemental targets in his vacuum tubes varied in a simple,
linear way with atomic number. The K-α line, which because it is the
most prominent line dominated his measurements, varies in frequency

4

approximately as:

$$2.48 \ 10^{15} \ (Z - 1)^2$$

which corresponds to energy in keV varying as:

$$C \ (Z - 1)^2$$

since for all electromagnetic radiation, the wavelength, frequency and energy are related through Planck's constant and the speed of light. The constant C has the following approximate values. Figure 1.3 shows the agreement between actual energies for K- L- and M-α X-rays and this simple expression.

Shell	Value
K	1.042 E-2
L	1.494 E-3
M	3.446 E-4

This approximate expression is not adequate to accurately locate and identify peaks in spectra, and we shall later introduce somewhat more elaborate equations that will. However, the general trend of increasing energy with atomic number was recognized early on. Figure 1.4 shows a plot of the energy of many of the prominent X-ray emission lines as a function of atomic number.

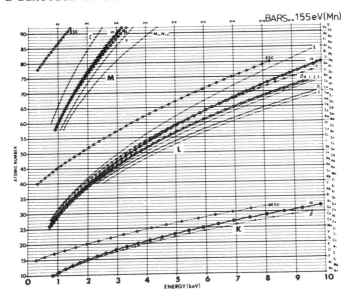

Figure 1.4 Energies of the principle emission lines of the elements (Fiori,1981a).

Excitation of an atom requires that enough energy be delivered by the incoming particle (electron, alpha, photon, etc.) to knock out the orbital electron. This energy is called the binding energy, or critical energy for the shell, abbreviated E_c. This value also increases with atomic number Z, and is widely tabulated and fitted to equations for simple calculations. The critical energy for a shell is always greater than the energy of the highest energy (shortest wavelength) X-ray photon emitted from it, since the electron that refills the shell comes from another (higher energy) bound level (which may later produce a lower energy X-ray itself). The excitation energy is the energy needed to raise the bound electron all the way to the conduction band. Since the outermost shells have very low binding energies (and produce X-rays with so little energy that they are not useful for analysis), the excitation energy is very close to the sum of the X-ray energies from the various combinations of transitions it can produce. For instance, for iron (Fe, Z=26), the critical energy for excitation is 7.111 keV. The K-α-1 emission energy is 6.403 keV, but this emission occurs when an L III electron drops into the K shell, leaving a vacancy there. This could be refilled by an M V electron, with the emission of an L-α-1 X-ray having an energy of 0.704 keV. These total 7.107 keV., or nearly the same as the original excitation energy. Any further transition to refill the M shell will make up the difference, but the resulting photon will be too low in energy to be measured using any of the techniques we shall be considering for analytical spectrometry.

The description presented so far has ignored the possibility that an excited atom could rearrange itself to get back to the ground state in any way except that associated with the emission of a X-ray photon. If this were true, it would indeed simplify much of our interpretation of X-ray analytical results (it would also put out of business the people who make and use spectrometers that depend on the other kind of signal). The competing process is called Auger (pronouced O-Jay) electron emission. It can be thought of as two steps, one in which the X-ray is produced when the high energy electron drops into the lower energy vacancy, and then another in which the X-ray is itself immediately re-absorbed to ionize another, lower energy shell as shown schematically in Figure 1.5.

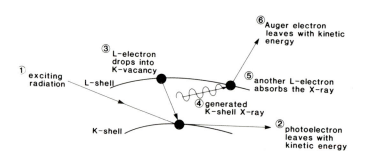

Auger electron production by internal capture of generated X-ray leaving behind two L-shell vacancies.

Figure 1.5 Schematic representation of the Auger effect.

6

Remember that we said before that excitation of an atom to create a vacancy could be caused by X-ray photons. The inner shell X-ray has too little energy to excite the shell from which it came, but enough to create a vacancy in an outer shell. In knocking out the bound electron there, the excess energy above that binding energy goes to impart kinetic energy to the ejected Auger electron (it is the measurement of this energy that is also used as an analytical tool). This process is sometimes called internal conversion.

The relative probability of X-ray emission or Auger electron emission occuring is constant for a given atom and shell. It does not depend in any way on how the atom was excited in the first place. The fractional probability of X-ray emission is called the fluorescence yield, and written ω . It varies from element to element as indicated in Figure 1.6. The reason for the trend has to do with the probability of the X-ray being absorbed by the outer shell. The cross section for X-ray absorption is something we shall examine in some detail later on. Simply stated, it increases as the energy of the X-ray gets closer to the binding energy of the shell, so that less of the energy needs to go into kinetic energy production. The difference between the energy of an inner shell X-ray (eg. a K X-ray) and the binding energy of the next shell out (in this example the L shell) decreases for lower atomic number elements (see the earlier plot of these energies). Hence, as atomic number decreases, the probability of Auger emission increases and the X-ray fluorescence yield drops.

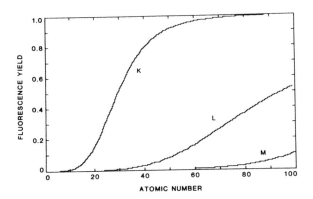

Figure 1.6 Fluorescence yield for K, L and M shell emission as a function of atomic number.

When the excitation of the atoms is produced by charged particles such as electrons or alphas, another process may also occur that causes emission of photons in the same energy range as the characteristic X-rays. There is no good English word for this process, which is called in German "Bremsstrahlung" (literally braking radiation). The result is to produce an additional portion of the measured spectrum in addition to the peaks from the characteristic emission lines, and that is called in English either continuum radiation or white radiation, because it spreads over many energies (or wavelengths or freqencies). Some spectroscopists refer to it as

7

background (it underlies the peaks) or noise (as opposed to the peaks, which are the "signal"). None of these is entirely a satisfactory name, and so the German word is perhaps the best one to use.

The discussion of excitation of atoms has thus far been concerned with interaction of the incident particle with the electron shells. The cross-section, or probability of this interaction occuring, is finite, and there is some probability that the particle will pass through the electron shells without interacting. If it does so, it finds itself moving through the region between the electron shells and the nucleus of the atom. Since the positive charge in the atom is concentrated in the nucleus and the negative charge in the electrons, it follows that this region has an electrostatic field. When a charged particle (eg. an electron) passes through an electrostatic field, it is deflected. This amounts to an acceleration, which takes energy and slows it down (reduces the kinetic energy). What happens to the energy? It is emitted as a photon. Since it comes about through the loss of speed ("braking") of the moving charged particle, it is called Bremsstrahlung, and since the energies of the photons are in the same range as the characteristic emission lines, we refer to them as X-rays.

The energy of each particular photon generated in this process does not depend on the element at all, but only on the geometry of the encounter and how close to the nucleus the trajectory passes. In the limit, the moving electron could lose all of its energy and be brought to rest if it were aimed directly at the nucleus. That is of course not likely. Other, lower amounts of energy are lost when the distance is greater (and the field strength is lower). From this reasoning, Kramers (1923) derived an approximate expression for the shape of the energy distribution of this radiation produced by an electron beam striking a solid target. The intensity $B(E)$ as a function of energy E is:

$$B(E) = kZ (E_o - E) / E$$

where k is a constant (Kramers' constant, about 2×10^{-9}, which as we shall discover much later is not really quite constant), Z is the atomic number of the target element, and E_o is the energy of the incident electrons. The effect of atomic number is not to change the shape of the distribution, but simply to increase the total amount of continuum in proportion to the total amount of charge in the atom, which governs the field strength. Consequently, we shall expect to find much more Bremsstrahlung background in spectra measured from high atomic number materials, and less from low Z specimens such as organic samples. Actually in the latter case, the continuum generated in the sample may be so low that it is dominated by other sources of spectrum background, including scattered electrons or continuum radiation generated by scattered electrons striking other (usually higher Z) materials near the sample such as holders, stages, pole pieces and so forth. This is an important source of difficulty we shall encounter in dealing with spectra from transmission electron microscopes, in particular.

From this expression, we would expect the continuum in the

spectrum to rise without bound as the energy E approached zero. This is not the observed shape of measured spectra, as indicated in Figure 1.7, and the difference is due to the absorption of the low energy Bremsstrahlung within the sample itself.

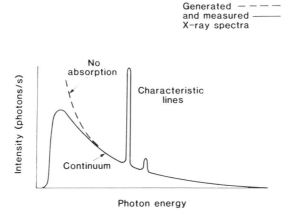

Figure 1.7 Schematic representation of a typical electron excited X-ray spectrum, as generated (dashed line) and as measured after leaving the sample.

Photon excitation does not produce Bremstrahlung radiation, because photons carry no charge. There is a small amount generated by the deceleration of photoelectrons excited from the atoms in the sample (and given kinetic energy) by the absorption of the exciting photons, but this is substantially less in amount than that produced by direct electron excitation, and also is produced deeper in the sample because the exciting X-radiation penetrates deeper, so that absorption of the continuum that is produced reduces it further. This abscence of continuum background is the major difference in overall appearance of electron and X-ray excited spectra. Excitation using heavy charged particles such as protons or alpha particles does produce Bremsstrahlung, but the heavier particles are not so easily deflected in the nuclear field and the energies and amount of continuum photons are much less than for electrons.

Chapter 2

Wavelength Dispersive Spectrometers

Bragg Diffraction

X-ray detection and measurement (to identify the emitting element) is generally accomplished by one of two types of X-ray spectrometer: energy dispersive (ED) or wavelength dispersive (WD). Although the ED type is now the more common, we shall discuss WD spectrometry first, because it came first historically, and because it offers a number of insights and points of comparison that will be useful later on. Both kinds of spectrometers give the same information: a spectrum of intensity (number of counts, or counts per unit time) versus an axis of either energy or wavelength of the X-rays. These are not independent quantities, of course, but simply two different ways to measure the same thing. The wavelength λ and energy E are related by

$$\lambda (\text{Å}) = 12.398/ E \text{ (keV.)}$$

It is because the method of operation of each spectrometer type is more directly described in terms of either wavelength or energy, and because the horizontal axis of the spectrum is most conveniently labelled in terms of those quantities, that the particular names have stuck on them. Wavelength, or diffractive X-ray spectrometers function by separating the X-rays of interest from the entire possible spectrum using diffraction from a crystal. This follows the Bragg equation:

$$n \lambda = 2 d \sin(\theta)$$

where n is any integer (although n=1 is by far the most commonly used condition, giving the greatest intensity), d is the interplanar spacing of the atomic layers in the crystal parallel to its surface,

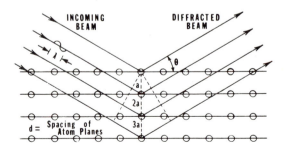

Figure 2.1 The extra path length of the wave diffracted from each successive plane of atoms is 2a = 2 d sin(θ). This must equal an integral number of wavelengths.

and θ is the angle of incidence of the X-rays. The sketch in Figure 2.1 shows the geometry for this phenomenon: when the extra path length for X-ray photons travelling to each successively deeper atomic layer, and the re-radiated waves from those atoms, equal an integral multiple of the wavelength, constructive interference occurs. The consequence is that a strong beam of diffracted radiation can be observed at the Bragg angle.

Since practical limits on the angle θ are roughly 20 to 75 degrees (set by mechanical design limitations in the spectrometer), a particular crystal (whose atomic spacing is commonly described by the 2d value for convenience) can cover a wavelength range of about 0.5 to 2 times d. Since crystals are available with 2d values from 1 to 10 angstroms, and on up to 60 angstroms in special cases (psuedo-crystals made from layers of fatty acids), the range of coverage for X-ray analysis is about 1 to 100 angstroms (12 to 0.1 keV.). This covers the K, L, or M lines of elements from Be (Z=4) up in the periodic table.

Since a perfect crystal would give peaks only a few arc-seconds wide, it is more common to use mosaic crystals which produce broader peaks, easier to locate with real, mechanical spectrometers. These also give greater total reflectivity (5-20% instead of <1%). Sometimes physical abrasion of a too-perfect crystal will produce the desired results. For long wavelengths, ruled gratings can also be used for diffraction. This is particularly useful for measuring shifts in wavelength or intensity of outer shell lines, for chemical bonding studies. This cannot be done with energy-dispersive systems.

Spectrometer Mechanisms

In the case of electron beam excitation (the SEM, TEM or microprobe), the source of X-rays is essentially a point, as far as the dimensions of the X-ray spectrometer are concerned. A flat crystal, as shown in the sketch of the Bragg condition, would only diffract along one line across the surface. The ideal case for a point source is achieved by bending and grinding a crystal in two directions, but practically this is not possible. Johannsen focussing as shown in the sketch of Figure 2.2, involves bending the crystal to

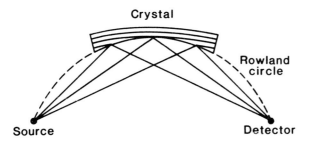

Figure 2.2 Focussing arrangement of source, diffracting crystal and detector collects X-rays emitted over a range of angles from the point source.

radius 2R and grinding the surface away to radius R (the radius of the 'Rowland circle' that defines the spectrometer geometry). This permits diffraction from the entire crystal, but for soft crystal materials is not practical (these would be bent but not ground, giving less perfect focus and lower efficiency).

The spectrometer mechanism that moves the detector (still to be discussed) and crystal through a range of positions and Bragg angles, to survey a range of different X-ray wavelengths (energies), can be of two principal types: 1) rotating, in which gearing is used to move the detector through twice the angular distance as the crystal turns; or 2) linear, in which the crystal, detector, and Rowland circle all move as shown in the sketch below. Geared or rotating mechanisms are commonly used in X-ray fluorescence spectrometers (using photon excitation from an X-ray tube). The linear design is far more complicated, but is now the most common in electron microprobes because it keeps the same solid angle and same takeoff angle for detected X-rays. Mechanisms to change crystals, and to position slits in front of the detector to collimate the beam, add to the complexity of these spectrometers. The attachment of motors with suitable control (often a small computer) permits the efficient sequential positioning of the detector and crystal at angles which correspond to the diffracted wavelengths of characteristic X-ray lines from a series of elements of interest.

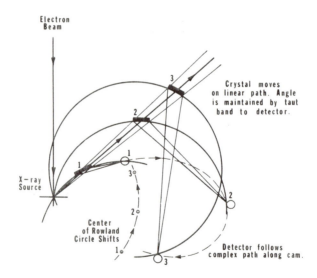

Figure 2.3 Linear spectrometer mechanism moves the crystal along a straight line path, intercepting a constant solid angle of X-rays. Distance from crystal to source is proportional to the diffracted wavelength.

Either mechanism can be oriented upright or sideways. Note that moving the source by as little as 10 to 100 micrometers will defocus the X-ray optics and reduce the intensity. For a horizontal spectrometer (Rowland circle parallel to the floor), the worst defocus case is moving the beam horizontally across the specimen, toward the X-ray spectrometer, while up-and down-displacements from a somewhat rough surface cause less of a problem as the diffracting loci move across the width of the crystal. For the vertical spectrometer orientation, this dependency is reversed.

Crystals

Several crystals are needed for typical analysis. Some common crystals are:

LiF	(200)	lithium flouride	$2d=4.027\,\text{Å}$
SiO_2	(1011)	quartz	$2d=6.686\,\text{Å}$
PG	(002)	pyrolitic graphite	$2d=6.71\,\text{Å}$
PET	(002)	penta-erythritol	$2d=8.742\,\text{Å}$
ADP	(101)	NH_6PO_3	$2d=10.64\,\text{Å}$
KAP	(1010)	acid phthalate*	$2d=26+\text{Å}$
STE		stearates**	$2d=60-150\,\text{Å}$

*K, Rb and other elements are the cations
**lead stearates, lignocerates
and other fatty acid films

The soap film 'crystals' for high 2d work have a short life. They are made by spreading a fatty acid (eg. stearic acid in benzene) on water and then picking up monolayers by dipping and withdrawing a substrate. The hydrophilic and hydrophobic ends of the molecules cause them to line up in parallel arrays, so that layers a single molecule thick adhere to the substrate (or each other) on subsequent passes through the surface. The presence of heavy atoms such as lead provides sufficient X-ray scattering for acceptable diffracted intensity from these otherwise organic psuedo-crystals, although intensities are still low (a few percent). Some of the other crystals also have problems. PET deteriorates in vacuum, for instance, and has a high thermal expansion coefficient (at θ = 45 degrees a 10°C temperature change will shift a peak by about .2 degrees). Another less common approach is to use a single crystal of mica (2d=19.84 angstroms) and cover the entire 1-15 angstrom range by using various 'order' (values of the integer n in the Bragg equation) diffraction lines (the high order lines still have good intensity with mica, but the spectrum becomes quite cluttered). In most systems, pulse height discrimination is used to reject high order lines. This also reduces spectrum background due to high order scattering of continuum radiation, which is particularly significant with high atomic number samples.

Proportional Counters

Several types of X-ray counters are used, but we shall here restrict ourselves to gas proportional counters, as being the most common in microprobe work. Figure 2.4 shows the essential features. The entrance window is generally an organic film such as collodion, mylar, polypropylene, etc., sometimes supported by a metal grid for strength. It must be periodically replaced. The thinner windows used for light element work are permeable to the gases and continuous flow of gas to the counter is required. The window is coated with a thin layer of evaporated aluminum to provide electrical conductivity.

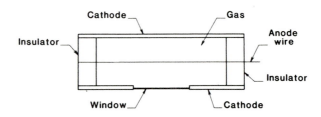

Figure 2.4 The gas filled proportional counter may be either sealed or have a continuous flow of gas through it.

X-rays ionize the gas atoms, leaving electrons free to move to the central anode wire and the positive ions to the case. Typical gases require 25-30 eV for ionization, so a 1 keV (12 angstroms, eg. Na K line) photon would produce 35 electron-ion pairs while a 10 keV (1 angstrom, eg. Pb L line) photon would produce 350. Neither signal would by itself be large enough to be easily detected, measured or counted. We rely instead on gas amplification: the voltage on the detector is raised high enough that the electron and ions are accelerated until their scattering from other atoms in the gas causes further ion pairs to be formed, producing a bigger signal. The critical voltage depends on the wire radius and the gas being used. In any case, gas gain takes place only within a few tens of micrometers of the central anode wire, where the field strength is high. The gain G is given by:

$$\log G = (AV/\log(r_2/r_1))\log(V/BPr_1\log(r_2/r_1))$$

where A and B are constants, P is the gas pressure, V the voltage, and r_1 and r_2 are the radii of the anode wire and counter wall, respectively. From this equation, the sensitivity of the gain to pressure and voltage is apparent. Gain stability of 1% requires voltages to be stable to one part in 10,000.

As the voltage is increased on a detector of this kind, its behavior changes. Initially, it acts as a simple ionization chamber with no gas gain. Then there is a fairly broad region where stable gas gain, roughly proportional to the voltage as described by the gain equation above, it observed. This is called the "plateau". Then as the voltage is further increased, the region where the gas gain is taking place increases in volume and length along the wire until it involves the entire length of the detector. When this breakdown occurs, a very large output pulse is produced, but no longer has any proportionality to the photon energy, and in addition the time for the breakdown to occur and all of the ions to migrate to the outer case so that the detector can produce a distinguishable new pulse in response to a new entering photon is very long - often hundreds of microseconds. The detector has become a Geiger counter.

Electronic amplification of the pulses, discrimination to reject high order diffracted lines (by comparison of pulse heights to preset voltage levels), and counting to determine intensities, all use conventional electronic modules. The data may be presented as a strip-chart recording of count rate versus time, as the angle θ is varied by a motor. For quantitative analysis, counting at preset angles for preset times gives the elemental intensities.

Even the normal proportional type of counter is 'dead' for the length of time needed to move the electrons to the wire, and will show reduced gain until the ions reach the outer wall because they shield the electrostatic field. The dead time T is given by:

$$T = r_1{}^2 \log (r_2/r_1) / 2 \, V \, k$$

where $k = 1.6 \text{ cm}^2/V/\text{sec}$ for 'P-10' gas (argon + 10% methane). The output pulse after amplification is about 1 microsecond in duration, and with a height roughly proportional to photon energy. Pulse height discrimination is used to reject counts from X-rays that diffract with higher values of n. The distribution of pulse heights is broad because the number of ion-electron pairs formed is not constant, but has a roughly Gaussian (actually Poisson) distribution that arises from the statistical processes involved in ionization and scattering. From 'normal' counting processes, we would expect the uncertainty in the number of events to be the square root of the number ($\sigma = \sqrt{(N)}$). However, in this case there are multiple pathways for energy loss, and the uncertainty is reduced to $C = \sqrt{(FN)}$ where F is the Fano factor, about 0.8 for a gas counter. Hence the resolution (full width of the peak at half its maximum height) is about 15% for an iron K peak.

The distribution also shows an escape peak 3 keV. below the main peak (for argon gas) 5-10% of its height, due to Ar K X-rays excited in the gas that leave the detector (the gas is relatively transparent to X-rays, so the probability of escape is high). This phenomenon will be encountered again in dealing with energy-dispersive spectrometers. For Fe, the escape peak is near K (potassium), but of course it is measured when the spectrometer θ angle does not correspond to potassium. There are a few cases for high values of n in which escape peaks can overlap or be confused with diffracted X-rays from elements in the sample.

System dead time is the total time during which one pulse is detected, amplified, and measured, and another one (another X-ray) could not be distinguished as a separate event. It depends on the pulse width, and is typically of the order of a few microseconds. Since the X-rays do not enter the spectrometer in a regular sequence, but randomly in time, the relationship between the true count rate and the observed rate is:

$$R_{true} = R_{obs} / (1 - R_{obs} (\tau))$$

where τ is the dead time, and is nearly constant (there is some dependence on photon energy) for a given counter and electronics. For a typical dead time of 2.5 microseconds, at a true input count rate of 50,000 pulses per second, the measured rate would be 44,400 cps, a loss of just over 11%. For lower count rates, the loss often becomes negligible. At very high count rates, the gas counter shows a drop in pulse height due to the drop in effective V from electrostatic shielding by positive ions. Pulse height is also affected by impurities in the gas or dirt on the anode wire.

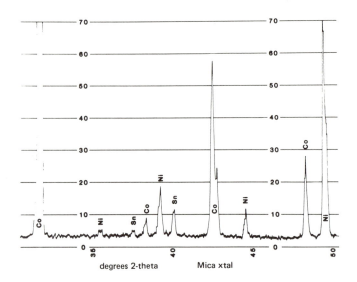

Figure 2.5 Typical wavelength dispersive scan from an electron microprobe.

Summary of Characteristics

Wavelength dispersive systems offer high resolution (sharp peaks in plots of intensity versus angle) that easily separate the emission lines from different elements. They also have a good ability to process high count rates. However, the diffracting process has very low efficiency, and the restrictive X-ray optics intercept only a small solid angle of the emitted radiation. The spectrometer signal is very sensitive to positioning of the source, or sample surface. The spectrometer must be set mechanically for each element of interest, and so may miss elements that are not anticipated. The mechanisms are complex, expensive, and somewhat difficult to automate, service or maintain. In the next section, we shall consider the now more common energy dispersive systems, and will see that while they offer a different set of advantages, they do not excel in all respects.

Chapter 3

Energy Dispersive Spectrometers

History

The energy dispersive (ED) spectrometer first gained acceptance as an attachment for the scanning electron microscope (SEM), principally because the already well established WD types were not useable with the SEM because of their low detection efficiency. The diffraction process is inherently inefficient, reducing the intensity by an order of magnitude or more in most cases, and the resrictive X-ray optics accept only a small fraction of the emitted photons (solid angles for all but a few, highly specialized spectrometers are about 0.01-0.001 steradians). By comparison, the ED detector can be positioned close to the specimen (even in SEM's that have no clear access path from a large port positioned at a high takeoff angle from the specimen, as would be needed by a WD spectrometer), to cover as much as 0.1 steradian. Except for minor losses (under 5%) due to scattering and escape events, every photon that reaches the detector produces a measurable, countable pulse. At very low energies (below 2 keV.), absorption in the entrance window reduces the efficiency, and at high energy (above about 20 keV.), penetration through the detector thickness can also cause efficiency to drop, but even so, total efficiencies as much as three orders of magnitude greater than WD systems are possible, and allow the ED spectrometer to function at the low beam currents used to obtain high resolution SEM images. There is also no difficulty in detecting X-rays from the rough sample surfaces normally encountered in the SEM, since no restrictive diffracting geometry is involved. This is not to imply that rough surfaces do not alter the elemental X-ray intensities, in ways that often frustrate quantitative analysis, and we shall return to this problem later on.

The first use of ED spectrometers was for completely qualitative analysis, that is, the identification of the elements present at the point on the sample surface being struck by the electron beam. The fact that all X-rays entering the detector, with many different energies, could be 'simultaneously' (at least with respect to the human operator's time sense) measured, to accumulate an entire spectrum showing even unexpected elements, was lauded as a qualitative analysis aid. The comparatively poor resolution (wide peaks in the spectrum) made it difficult to detect minor or trace elements (they could not be distinguished above background), and hid the peaks from some elements where they were overlapped by lines from others. The resolution of these detectors, initially over 500 electron volts full width at half maximum height (FWHM), improved quite rapidly as many manufacturers attacked the problems in their effort to satisfy this new, burgeoning market. As fundamental limitations were reached, this development has levelled off. The resolution of ED systems, while it may improve slightly in the future, will never approach that possible with the WD spectrometer. Typical peak widths for Mn K α , used by all manufacturers as a convenient specification (since the X-rays can be obtained from a radioactive source of Fe 55, needing no electron beam for testing purposes), are now under 150 eV., or about 2.5% of the peak energy.

We shall be examining the principles of operation of these systems, to understand these limitations (and others), and to see how to respect, circumvent or overcome them.

Principles of Operation

Figure 3.1 illustrates schematically a typical ED detector. It is a lithium-drifted silicon "Si(Li)" crystal, a few millimeters thick and under a centimeter in diameter. Other energy dispersive detectors include either lithium-drifted or ultra high purity germanium (used primarily at much higher energies -in the MeV. range), or mercuric iodide (operable at room temperature with somewhat poorer resolution than Si(Li)). The description that follows is substantially correct for those devices as well.

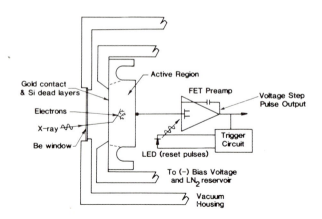

Figure 3.1 Diagram of typical Si(Li) detector.

When an X-ray photon enters the body of the detector, it is absorbed to create a high energy photoelectron. This electron strikes other nearby atoms in the crystal, losing energy itself but ionizing other atoms and freeing more electrons. As a statistical average, the number of electrons finally liberated, and raised above the band gap in this semiconductor material, is given by:

$$N_{electrons} = \text{X-ray energy (eV.)} / 3.81$$

where 3.81 is the average energy for electron-hole pair formation at the normal operating temperature of the detector, about 80-90K. Other materials have different constants: about 3 eV. for germanium and 4.2

eV. for HgI_2 . For a 1 keV. X-ray (eg. Na K), this means there are only about 260 electrons free in the detector, and less than 2000 for a nickel K X-ray (7.47 keV.). The signal we are trying to measure is very small, and the solid state detector is used as an ionization chamber without the 'gas gain' described for the gas proportional detector used in WD systems.

The freed electrons (and holes) drift under the influence of the applied electric field (the bias voltage), the electrons going to the rear of the detector and the holes to the negatively charged front surface. If all of the charge makes it to the electrode without being trapped in lattice imperfections (the lithium atoms are diffused into the crystal to electrically compensate any imperfections that remain after careful zone refining to create perfect single crystals) or recombining, the total charge is proportional to the original photon energy, which we want to measure. The bias voltage is set high enough to move the electrons and holes, which have roughly equal mobility in silicon, out of the device in less time than the recombination time. This requires a few hundred volts per millimeter of thickness. Too high a bias voltage increases the leakage current and degrades resolution.

If the carrier mobility varies, as it does in the surface layers of these devices (either due to depletion of the lithium by diffusion, or the creation of surface electronic states), it can slow down the arrival of some of the charge. This would reduce the size of the induced current pulse, resulting in a measurement of photon energy that is too small. Generally, physical collimation of the detector is used to prevent X-rays from entering near the edges of the active volume (various electrical collimation schemes, referred to as guard rings, have been devised as well, but they add considerably to the complexity and size of the devices, and hence to the cost). The front and rear surfaces are harder to protect. Very high energy photons that interact near the rear of the detector are generally of less concern in SEM applications than low energy ones that are absorbed near the front. These latter X-rays come from relatively light elements (or outer shells of heavier elements, but for them there is usually another higher energy line that can also be detected), and the charge straggling that occurs results in some pulses being a bit too small. The stored peak comprises counts from many X-rays, and would otherwise be symmetrical, but becomes distorted with a low energy tail containing the counts from X-rays whose measured pulse was thus reduced in size. This can contain 5 -10% of the counts in the peak, and can introduce great complexity and increased errors in trying to fit and deconvolute peaks and model background in the low energy range. The phenomenon is usually referred to as part of the effect of the silicon 'dead layer', meaning the surface layer of the detector that is truly inactive electrically and simply represents an additional entrance window through which the incident photons must pass, but actually the effect is more serious and subtle than simple loss by absorption of some X-rays.

We shall return much later to a discussion of the absorption of X-rays in the entrance window, made up, as we see, of the beryllium or organic material used to separate the vacuum system of the detector cryostat from that of the microscope, as well as the very thin layer of evaporated gold used as an electrical contact on the front face of the detector, and the inactive or 'dead' layer of silicon under it which is actually a very thin "P" region in what amounts to a P-I-N semiconductor diode with a large intrinsic region. We shall now continue following the electrons and our efforts to count them.

As in the gas proportional detector, the number of electrons produced by an X-ray of given energy is not constant, but varies in a statistical way. The standard deviation of the Poisson distribution for the number is σ = square root (N F) where, as before, F is the Fano factor, describing the way in which energy is lost in competing processes, to the ionization of atoms and excitation of vibration states in the lattice. It is the 'correlation' between subsequent events in the life-history of individual scattering electrons that results in better resolution (less spread in the number of electrons) than simple statistics would lead us to expect, and it is the presence of a solid crystal lattice with few vibration states that produces a much lower Fano factor for these detectors than for the gas counter. The Fano factor is about 0.11 for Si(Li) detectors, about 0.13 for germanium, and above 0.2 for mercuric iodide (Dabrowski, 1982). These numbers vary somewhat for individual devices, depending on the quality of the material and the fabrication methods.

Preamplification

The electron charge from the detector's rear contact enters a field-effect transistor (FET) used as a first stage amplifier. This integrates the charge and produces at its output a small voltage step, with a rise time comparable to the charge collection time (very short, typically 10-100 nanoseconds). The output of this amplifier stage includes electrical noise, however, resulting from the capacitance of the detector itself and any other connections, including the feedback resistor needed to restore the output voltage between pulses. In most modern systems, the feedback resistor has been eliminated, and the preamplifier output voltage increases in a ragged staircase as indicated below, until it reaches a level at which non-linearity in output could occur (a few volts). Then another circuit is triggered which pulses a light-emitting diode in close physical proximity to the junction of the FET (whose encapsulation has also been removed to reduce capacitance and allow the light to reach the junction). This light creates photoelectrons in the FET itself which short it out, restore the output voltage to zero, and allow the process to continue (Landis, 1970). The time needed for this 'pulsed optical feedback' reset pulse is typically 50-100 microseconds (only 10-20 microseconds for the light pulse, and the rest for the transients to die out), during which time a few X-rays may be missed (the subject of dead time and its correction comes later).

The detector and the first stage of the amplifier are cooled to low temperature using liquid nitrogen, to reduce the dark current flowing through the detector in the absence of X-rays (due to thermal electrons above the band gap). The optimum temperature for the FET, at which it has the lowest electrical noise characteristics, is typically

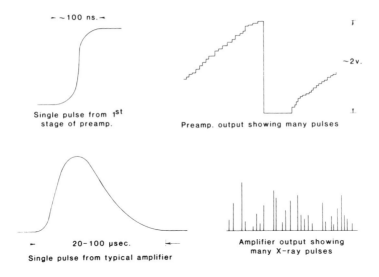

Single pulse from 1st
stage of preamp.

Preamp. output showing many pulses

~2v.

~100 ns.

Single pulse from typical amplifier

20-100 µsec.

Amplifier output showing
many X-ray pulses

Figure 3.2 Signals present at various stages in the energy dispersive amplification process.

about 120-140K. It may be warmed slightly relative to the silicon detector by a small heater or by the design of the cold finger from the LN reservoir used for cooling. There is some additional dark current (the total is generally under 1 picoampere) flowing in surface states around the edges of the detector, but special geometries ('ditches', 'top-hat' shapes, etc.) have been devised to reduce this contribution. The recent developments in room temperature detectors such as mercuric iodide have been made practical by the low leakage currents (1-10 picoamperes) obtainable without cooling.

Pulse Processing

The step-shaped pulses from the preamplifier must be further amplified and shaped by integration and differentiation before they can be measured. The conflicting requirements placed on the pulse processor include: 1) very low noise amplification, which is mainly achieved by using long integration time constants to average out noise ripple, and 2) high throughput capability to handle high count rates with minimum loss or distortion of pulses due to overlap, or shifting of pulse heights due to baseline (zero voltage) drifts. In addition, high long term stability and excellent linearity are essential to obtaining useful final calibration at the multichannel analyzer display.

Most amplifiers use a 'semi-Gaussian' pulse shaper, with integration time constants of several (2-16) microseconds, giving total pulse widths before the voltage has returned to the baseline, of tens of microseconds (40-60 is not uncommon). Note that these times

21

are enormously greater than the charge collection time in the detector itself. X-rays that enter the detector a microsecond apart are totally separate events, as far as the detector and FET are concerned, but the signals will be added together in the amplifier creating a single pulse with a magnitude equal to the sum of the separate original signals. This phenomenon, known as pulse pile-up, clearly becomes increasingly important as the count rate (number of X-ray pulses per second) increases. It is pulse pile-up in the amplifier that ultimately limits the count rate capability of ED systems, and some ingenious solutions have been proposed to at least partially overcome the limitations.

The first correction applied to the counting process was to correct for dead time in the pulse processor. In most ED systems, the multichannel analyzer incorporates provisions for counting for a preset time. Since there will be some loss of counts due to the dead time of the system, usually the time that is preset is interpreted as the live time rather than clock or elapsed time. Live time can be thought of as the difference between clock time and dead time, or as the time when the system is quiescent (not busy processing a pulse) and able to respond to one that comes into the detector. Early dead time corrections functioned by stopping the system clock while the multichannel analyzer itself was busy, and this was soon extended to allow the amplifier to stop it as well during the time spent processing a pulse (to do this, of course, means that the amplifier must be able to recognize that it has a pulse, either by monitoring the input voltage and seeing a rise above a certain threshold, or by having a separate amplifier with a short time constant to produce a 'fast' pulse -a technique we shall take up next). Stopping the clock for preamplifier resets is also incorporated in most modern systems.

Pulse pile-up rejection is intended to prevent the measurement and counting of pulses that are interfered with by another pulse too close in time (before or after) to permit a complete return to the baseline voltage between pulses. It also corrects for the time lost when these pulses are rejected. The method used requires a second amplifier, usually with a much shorter integration time constant (under 1 microsecond). This produces output pulses whose height is subject to considerable variation due to electronic noise, which has been inadequately integrated out, but which occur very quickly after the arrival of the X-ray at the detector, and which have a much greater chance of not interfering seriously with each other, even for events which have pulses that are completely added together in the main amplifier with its longer time constant. The 'fast' pulses trigger a timing circuit that checks to see whether two consecutive pulses are far enough apart to allow accurate final measurement. If they are not, one or both (depending on their distance apart, since the pulses are not symmetric in time but have a faster rise time and slower decay time, so that pulse number 1 might be measurable and safely past its peak before pulse number 2 arrives, but pulse number 2 might be riding on the slowly decaying tail of the first pulse and not measurable) are gated off, that is, an electronic relay prevents them from going to the multichannel analyzer for measurement. Figure 3.3 shows the time relationships between fast and slow pulses for different pile-up conditions.

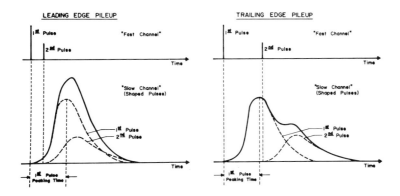

Figure 3.3 Pulse pileup in the amplifier may involve either distortion of both pulses (leading edge pileup) or just the second pulse (trailing edge pileup) (Woldseth,1973).

Counting Time Corrections

When this rejection takes place, it is also necessary to correct the system clock for the lost time. Sometimes this is done by stopping the clock for the length of time between the pulses that piled up, or for the processing time of the amplifier. Another, more elegant solution is to keep track of the number of rejected pulses in a small counter, and from time to time when this reaches a predetermined value (perhaps as often as every 16 or 32 counts), stop the system clock until that many measurable pulses have been accepted and stored, and the counter counted back down to zero. This method is very simple to implement in construction, and is exact in its correction provided that the pulses in the 'fast' channel are not piled up. But the shorter time constant used there (to time-resolve pulses very close together) gives poorer energy proportionality, and increases the chance that pulses from low energy X-rays, which are smaller to begin with, will be indistiguishable from the noisy baseline level and may not trigger the pulse detection circuit by rising above the voltage discriminator threshold. In other words, it is harder to properly reject pile-up pulses at low energy than at high energy, and depends much more on the proper adjustment of the pile-up discriminator. Some systems even use two 'fast' channels with different time constants, one for low energies (below about 2 keV.), and one for higher energies. None of the schemes works adequately for the ultra-low energy region below 1 keV., and consequently counting in that range can only be accurately carried out a very low count rates (below 1000 cps) where pile up is statistically very unlikely and ignorable.

A new development (Statham, 1982) in the effort to handle high count rates (at least on the scale of energy dispersive systems–they are not high as compared to WD specrometer capabilities) is the introduction of amplifiers with 'time variant' circuits. These aim to change the time constant after the peak of the pulse has passed, so

23

that a long integration time can be used to obtain a low noise peak for good measurement and ultimately good resolution in the spectrum, but then a short time constant can be used to quickly 'dump' the voltage and restore the output to the baseline. This can shorten the total pulse width dramatically, as shown in Figure 3.4. It is recognition of the arrival of a pulse in the 'fast' channel that triggers a sequence of timed stages using different system time constants. Even with a 'protect' time after the pulse to assure baseline conditions before the next pulse can be measured, the total throughput of a system can be doubled using this scheme. The cost of the electronics increases too, of course, and there may be a greater need for intelligent user adjustment and/or service than with simpler systems.

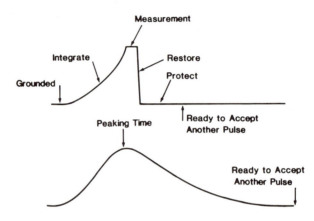

Figure 3.4 Comparison of semi—Gaussian 'normal' waveform with time variant or 'Kandiah' pulse shape. The latter permits long peaking times (for low noise due to integration effects) with short overall pulse width (Statham, 1981).

Still another approach to increasing the counting rate capability of ED systems is to 'de-randomize' the incoming stream of X-rays. Normally, X-ray production is random in time, so that for a given average count rate of R cps, the number counted in any time T will be 68% of the time within the range $TR \pm \sqrt{(TR)}$, 95% of the time within twice this standard deviation, and so on. If the source of excitation can be controlled (very rapidly) however, the spacing between photons can be made more regular. Although this is more commonly done with X-ray fluorescence systems, in which the X-ray tubes used for excitation are sometimes built with a control grid for this very purpose (Jaklevic, 1972), some SEM's and microprobes can be fitted with beam-blanking plates and circuits (intended for use in electron lithography for the manufacture of integrated circuits). Then the fast amplifier channel pulse can be used to shut off the electron beam for a period of time sufficient to process the pulse through the main amplifier, free from any possibility of pile-up (again assuming the fast channel resolves the events, and that no second X-ray was emitted before the beam could be shut off). When the beam is turned back on, the amplifier is ready to accept any new X-ray pulses from the detector.

By pulsing the beam on and off, a processed count rate up to three times greater than that obtainable with 'random' generation can be achieved, since virtually all detected X-rays can be counted. Of course, during the one-third of the time that it is turned on, the gun must operate at three times the power level, and must do so with no loss of stability. Few instruments can be practically equipped to operate in this way, but it is indicative of the importance of the counting rate limitation of ED systems that such heroic methods are employed.

For the conventional design of pulse processor, the total count rate performance as described by a throughput, or input-output count rate graph, shows the kind of behavior indicated in Figure 3.5. The various curves correspond to different amplifier time constants (which control the total pulse width). As practically all systems allow the user some choice in selecting the time constant, it is important to understand that the best system throughput (shortest time constant) should be used when it is needed to handle a high count rate situation, but that it is accompanied by poorer energy resolution because the shortened integration time averages out less of the noise. For many applications, particularly in the SEM, the count rates are not usually high. In the transmission or scanning transmission electron microscope (TEM, STEM), or in microprobe or X-ray fluorescence spectrometers, the count rate may be much higher, or may vary greatly (particularly a problem in the STEM), placing greater demands on the system throughput. The curves shown may not exactly describe a particular system, since not all manufacturers use exactly the same circuits for pulse integration, and in addition the curve

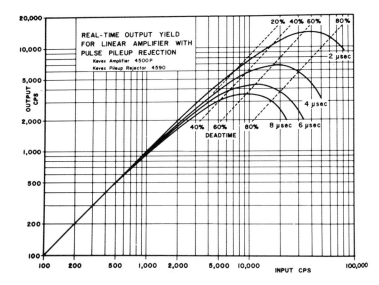

Figure 3.5 System throughput, or stored count rate capability, with normal amplifier pulse shapes and pulse pileup rejection shows maximum values that depend on the shaping time constant used (Woldseth, 1973).

25

depends not only on the number of pulses, but to some extent on their magnitude. The larger pulse heights that result from higher energy X-rays also produce greater pulse widths in the integration process, and it is really the energy product (sum of number of counts times their energies) that limits total system throughput. This also acts to the detriment of STEM applications, in which much higher energy X-rays can be produced than in most SEM's. The type of curves shown here are called "paralyzable", meaning that the system can be completely shut down by a sufficiently high input count rate. Some of the novel pulse processing schemes mentioned above are not paralyzable, so that a sufficiently high input count rate the output rate of pulses that can be measured and stored reaches an upper limit but does not decline.

In any high or even moderately high count rate situation, there is a significant amount of dead time in the system. By comparison to WD systems, where count rates over 10^5 cps can be handled with acceptable losses, few ED systems can be operated above 10^4 cps. Furthermore, of course, the ED system count rate includes all of the X-rays entering the spectrometer (at least those that pass through the entrance window), whereas the WD system need count only those X-rays of the element of interest, since the diffracting crystal has rejected X-rays of other wavelengths. At high count rates the system dead time for ED systems increases, so that a 100 second analysis (preset live time) may take much longer –perhaps 150 seconds of clock or elapsed time. Most manufacturers recommend that their systems not be operated above about 40% dead time, lest the accuracy of the dead time correction and the performance of the pulse processor (particularly the circuits that monitor the baseline voltage) be degraded.

In the corrections for system dead time, the key assumption is that the counts that are lost or rejected at one time can be made up by accepting other pulses later on, in other words that in any time interval within the total analysis time, the same energy distribution of incoming X-rays (within counting statistics) will be measured. For most situations this seems quite acceptable, but there are some important exceptions. For instance, if the sample is degraded by the electron beam (or other excitation), or if it is changing with time, we would be likely to realize that the analysis time interval would have to be kept short enough to resolve these changes. A much more insidious problem arises commonly in the SEM, but is likely to go unrecognized. Particularly in qualitative analysis, but also when trying to estimate relative concentrations of elements, many operators try to scan the beam in a raster over an area large enough to "average in" the small scale inhomogeneities they wish to ignore. However, these represent changes in composition which will produce different intensities of different energy X-rays. Similarly, changes in surface orientation as are found in many of the rough-surfaced specimens for which the SEM is considered the logical examination tool, will produce substantial changes in the X-ray intensities that reach the detector. Since it is not the average count rate that determines the throughput capability of the pulse processor, but the instantaneous count rate (time between successive photons), the average system dead time and the curves shown above can be quite misleading. If the X-ray intensity is very high when the scanning beam strikes a region containing a particular element, many of those pulses will be rejected. When the count rate is lower (the beam is on another region of the specimen surface), the system clock will stop while the lost pulses are "made up", but the element in question may not be in that region. Consequently the total analysis can be significantly biased against the element(s) present in the high count rate regions. The cases where this problem becomes most serious include the analysis of particles dispersed on a substrate (the sides of the particles that face the detector will give very high count rates, the substrate typically will not, and if most of the area scanned is substrate, the elements in the particles –especially the larger ones –may show artificially low intensities), and the analysis of trace but localized elements in organic materials (biologists take note). There is no easy solution to this problem, since keeping the total count rate low enough that substantial losses do not occur in the highest concentration, highest count rate region means that the ability to detect minor elements (limited by counting statistics) in a reasonable time will be poor. Systems in which the SEM beam is not continuously scanned in a raster, but advanced in steps on command from the ED system or directly positioned by it, can in principle overcome this problem since each point may be analyzed for a preset live time with its own dead time correction, but such systems are neither common nor inexpensive.

Resolution

The overall contribution of the amplifier electronics to the system resolution is usually as important or more so than that of the detector itself. If the photons entering the silicon detector were all of exactly the same energy, the pulses of charge reaching the FET would show a variation in magnitude in accordance with the statistics of free electron production as described before. This is usually called the statistical contribution to resolution, and varies as a function of energy since the number of electrons varies. The amplifier system, considered as a whole (and ignoring the details of the particular integration circuits each manufacturer has designed), also introduces a variation. If all of the pulses entering the amplifier were equal in size, the output pulses would also vary with a Gaussian or normal distribution because of the addition of random electrical noise. This "electronic" resolution is essentially energy independent. Since the two terms are independent of each other, they add together to produce the total system resolution:

$$\text{Res'n} = \sqrt{((\text{elect})^2 + (\text{stat})^2)}$$

Expressing the resolution as the FWHM (equal to 2.355 times the sigma for a Gaussian peak), incorporating the dependence of the statistical variation on energy (equal to 2.355 times square root (FE/3.81) where F is the Fano factor and 3.81 is the mean ionization energy in electron volts, as discussed before), and making use of the total system resolution measured for the Mn K α line, specified by most manufacturers, we can write:

$$\text{FWHM (eV.)} = (\ (\text{FWHM})_{\text{MnK}\alpha}{}^2 + 2.74\ (E - 5894)\ )^{1/2}$$

for the peak width expected for any energy E (in eV.). Figure 3.6

shows how peaks increase in width with energy following this relationship. The width at other fractions of the full height can be determined as simple multiples of the FWHM for the assumed ideal Gaussian shape; for instance the width at one tenth height (FWTM) should be 1.82 times FWHM.

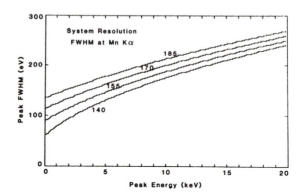

Figure 3.6 Variation of peak width with energy for a typical Si(Li) detector.

Spectral Artefacts

Although we have not yet discussed the counting and storage of the pulses to form the familiar spectrum displayed by the ED system, it is appropriate at this point to go over the various artefacts and imperfections that can occur in it, since they arise in the portions of the system electronics we have been describing. In Figure 3.7, a number of these artefacts are indicated, as they might result from a single monoenergetic beam of photons entering the spectrometer. Obviously, when a combination of many energies is coming in, the appearance of the spectrum is further complicated and some of the artefacts may not be easily seen; for instance the presence of spectrum background due to Bremsstrahlung produced by electron beam excitation will mask small escape peaks and partial pile-up regions. The principal peak is primarily an ideal Gaussian shape, whose width is given by the system resolution (as we have just seen due to the contributions of both detector and amplifier). On the low energy side, the tail that is shown results from slow charge collection in the detector, and will vary in magnitude and possibly in shape as a function of energy (and from device to device). Above the peak, at twice the energy, there is a small peak called the "sum peak" that represents two X-rays that entered the detector so close together in time that their pulses were added together in the amplifier and were not distinguished by the pile-up rejection circuitry. This peak can also occur when incoming X-rays of different energies are present, and sum peaks can be found at all possible combinations of main peak energies. The size of the sum peak is very count rate dependent, and at extremely high rates even triple or higher order sum events can be found. Below the sum peak is a broad region containing counts from pulses that were partially piled-up, with one pulse riding on the leading or trailing edge of another. Ideally, a perfect pile-up rejector should eliminate these partial pileups whenever the pulses are separated in time by enough that the fast channel pulses can be distinguished. In practice, this becomes more difficult at low energies where the fast channel pulses can hardly be distinguished from background noise levels, but the partial pile-up region will also appear whenever the discriminator setting for the fast channel pulse recognition circuit is misadjusted (and component aging, changes in system grounds, or interference from microscope electronics may also require readjustment of this setting).

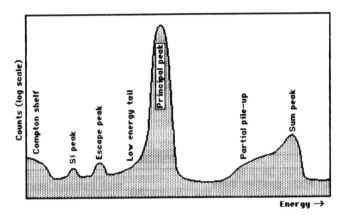

Figure 3.7 Schematic diagram of a single isolated peak as recorded by a 'real' energy dispersive spectrometer system. The various artefacts present are described in the text.

Below the main peak by exactly 1.74 keV. there is another, much smaller peak. This is the escape peak, just as for the gas counter. It results when the incoming X-ray or a high energy photoelectron ionizes a silicon atom in the detector by knocking out a K-shell electron, and the resulting X-ray photon is not reabsorbed in the detector (if it is, the energy is recaptured and contributes to producing the free electron pulse that is amplified and measured). Since X-ray absorption is much higher in the solid silicon detector than in the gas of the proportional counter, and since the fluorescence yield for silicon is lower than for argon (so that fewer X-rays are produced, with the excited atom being more likely to return to its ground state by an Auger transition with the emitted low energy electron simply making its own contribution to the free electrons in the detector), the escape probability is much lower for the silicon detector. For germanium and mercuric iodide, the higher energy characteristic X-rays from the elements in the detectors produce larger escape peaks which can be more troublesome in spectrum interpretation.

The ratio of size of the escape peak to the parent peak is a function of energy, which determines the average depth of penetration of the incident photon into the detector and consequently the probability of escape of the generated Si X-ray. It is useful to consider the general case in which the incident photons may enter the detector at an angle, because in many SEM's in particular, the range of positions for the specimen may cause this to occur. If the incidence angle is A (zero for normal incidence), the average depth of penetration is reduced by the cosine of A. Based on the mass absorption coefficent for X-rays of energy E in silicon (μ^E_{Si}), the depth distribution of the excited silicon atoms in terms of perpendicular distance into the detector z and the silicon density ρ_{Si} (and consequently the distribution of generated X-rays) becomes:

$$\mu^E_{Si} \; \rho_{Si} \; \sec(A) \quad \exp(-\mu^E_{Si} \; \rho_{Si} \; z \; \sec(A))$$

and the fraction of these which escape is:

$$0.5\omega \; (r-1)/r \; (1- \mu^{Si}_{Si} / \mu^E_{Si}) \; \cos(A) \; \log(1+ \mu^E_{Si}/(\mu^{Si}_{Si} \cos(A)))$$

The factor of one-half comes from the fact that for ionizations near the front surface of the detector, half of the photons are emitted toward the surface and half into the body of the detector. The constants ω (fluorescence yield= 0.043 for Si), r (the K absorption edge jump ratio= 13 for Si), and expressions for the silicon absorption coefficients for energy E and for its own K X-rays (327.9) which we shall develop in later chapters, have been combined by Statham (1976) into a single expression used to numerically calculate the escape peak fraction:

$$0.0198 - 0.000328 \; E^{2.77} \; \cos(A) \; \log(1 + 60.24/(E^{2.77}\cos(A)))$$

from which the height of the escape peak relative to the parent peak becomes calculable as:

$$0.0202 / (1 + (mE+b) \; E^2)$$

where $\quad$ m = 0.01517 cos (A) −0.000803

and $\quad$ b = 0.0455 cos (A) + 0.1238

This expression is quite readily incorporated into computer programs that identify or correct for escape peaks. Generally, escape peaks increase in magnitude with oblique incidence, and also increase as the energy of the incident photons is reduced, down to about 1.9 keV. (below the Si K absorption edge they can no longer excite the Si K shell, of course). Figure 3.8 shows the relative height of escape to parent peak for several incidence angles.

Also shown in the "artefact" spectrum is a small peak just at 1.74 keV. This represents Si K X-rays that are fluoresced from the dead silicon layer on the front surface of the detector. Because of the proximity of this layer to the region in which slow charge collection takes place, and the possibility that the energy loss process may produce some free electrons in the dead region, reducing the magnitude of the collected charge pulse, this peak is very asymmetric with a

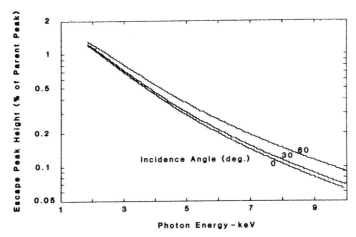

Figure 3.8 Ratio of escape to parent peak height as a function of energy and incidence angle of radiation at detector (0 degrees = normal incidence).

broad low energy tail. In most real spectra it occurs just at the discontinuity in the continuum background of the spectrum due to the absorption edge step of the silicon of the detector, and is consequently hard to distinguish and even harder to correct for.

Finally, at low energy there is rising background level, sometimes with a flat shelf, that makes ultra-light element analysis (X-rays below 1 keV.) difficult. It has several causes. When very high energy (20 keV. and above) X-rays enter the detector, the probability increases that they will lose energy by Compton scattering as well as by photoelectron production. If the Compton scattered photon leaves the detector, which is likely at these energies (the Compton cross section rises with energy, and is about one tenth that for photoelecton production at 25 keV in silicon), the small amount of energy deposited in the detector produces a count at very low energy in the spectrum. Low energy artefacts also come from stray light reaching the detector, or from microphony. The input to the FET is an extremely sensitive capacitative microphone, and any vibration transmitted through the mounting of the components will produce low energy pulses that are stored in the spectrum (they may in fact sometimes be at much higher energy if a particular resonant frequency is struck, or if internal damage occurs to the support structure).

The most important spectral artefact not shown in the figure, which is peculiar to electron excited spectra, is the response of the detector to electrons that penetrate through the entrance window. These produce a continuum of energy extending up to the maximum energy that can penetrate the window. This severely distorts the spectrum and makes efforts to fit the background mathematically using semi-theoretical expressions hopeless. The problem is worst, of course, with higher beam voltages and thinner windows. At a 25 keV. accelerating voltage, a common SEM operating condition, some penetration through the typical 7-8 micrometer beryllium window can occur. At 100 keV. in a STEM the problem would be so serious as to completely mask the X-ray spectrum, except for the fortuitous fact that the specimen in a STEM is usually positioned in a high field region of the lens, and scattered electrons can usually not travel in straight lines to enter the detector (sufficient shielding is needed to keep penetration through the spectrometer housing from occuring). With ultra-thin windows, or "windowless" (ie. no beryllium window −the gold and dead silicon layer are still present) detectors used for very light element work, magnetic or electrostatic electron traps are essential. They would be extremely useful in all systems capable of operation at energies above about 20keV., but only a handful of users have built their own, and the manufacturers of the X-ray spectrometers and microscopes seem equally disinclined to design or incorporate them. With excitation by photons (X-ray fluorescence analysis) this is not a problem.

Pulse Measurement

The amplified pulses are measured by an Analog to Digital Converter (ADC) and counted by a Multi-Channel Analyzer (MCA), both of which are often housed in the same cabinet and so may not be distinguished. The MCA may also, in fact, be the computer used later to process the data, but for our purposes here we can consider it to be a separate entity, at least in function.

Analog to Digital conversion of a pulse of the type coming from the amplifier requires a circuit to recognize the presence of a pulse and find its maximum, and another to convert the height of the pulse to a number. Although there are several types of ADC circuits used for general types of voltage measurement, the one selected almost universally for ED spectrometers is the Wilkinson or run-down type.

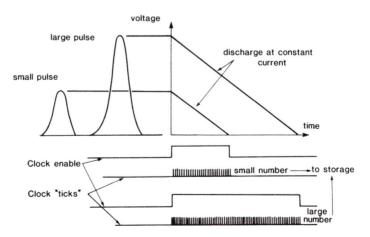

Figure 3.9 The Linear or Wilkinson ADC measures pulse height by the length of time (number of clock pulses) required to discharge a capacitor from the maximum pulse voltage.

32

Its mode of operation is summarized in Figure 3.9. The incoming pulse has its peak detected by the change in slope from positive to negative, and the voltage input is then disconnected, leaving the peak voltage on a small capacitor. This may be done simply by charging the capacitor through a diode, so that when the pulse voltage drops, the capacitor is left with the peak voltage. This is discharged using a constant current supply, while a high frequency (10 to 100 MHz) oscillator counts the time needed to return to the baseline voltage. It is the number of oscillations, or clock 'ticks', that becomes the number representing the pulse height, and is used as the address in the MCA to store the information. This is done simply by adding one to the contents of memory at that address, whether the memory is a special purpose one or part of the memory of a small computer.

Although not widely used in ED systems yet, another type of ADC holds promise or the future because it is much faster than the run-down type, and has a digitisation time independent of pulse magnitude. This is the "successive approximation" ADC. It functions by guessing at the height of the pulse, trapped as before in a holding circuit or capacitor. The first guess is just one half of the possible full scale magnitude, and is used to generate a voltage using a digital to analog converter that is compared to the actual pulse voltage. If the voltage is larger than the guess, the next guess will be three quarters of the maximum, and if it is less one quarter. One each successive guess, the size of the step is reduced by a factor of two. In 10 steps it is possible using this scheme (which is employed in some children's games) to find any number between 0 and 1023, which means that the pulse height can be measured to that precision and the ADC produces an address suitable for a 1024 channel memory (12 steps give one part in 4096). The number is recorded by setting successive bits (binary digits) to one (if the pulse height is larger than the guessed value) or to zero (if it is smaller). The string of ten or twelve bits forms a binary representation of the number which is the ADC output and the memory address. This type of ADC is commonly used in other applications as for the measurement of slowly varying voltages. Adapting it for pulse measurement and making sure that the successive steps are precisely in ratios of 2 (necessary to obtain a linear conversion performance) will make this type of ADC useful for ED systems in the near future.

The display of the information from the memory on a cathode ray tube, often scanned at television rates, can be performed using either analog circuits that produce voltages to deflect the beam, or digital techniques in which timing to turn the continuously raster-scanned beam off and on produces the viewed image (as in conventional television). Without intending to give unnecessarily short attention to the ADC/MCA end of the spectometer, which is both vital and expensive, it is less important for our present purpose since it has less opportunity to influence (distort) the nature of the stored data. The description of MCA operation which follows should be recognized as a diversion from our main thread, interesting as it helps our familiarity with the hardware we depend on, but probably only of real interest to those who will create their own instruments, as for instance by interfacing a microcomputer to an ADC and programming it to count pulses and display a spectrum.

33

Multichannel Analyzers

The ADC presents a number, the number of clock ticks from its high frequency oscillator, usually in the form of a binary value held in a register. The number of bits defines the magnitude of the number –12 bits for instance can express a number from 0 to 4095 while 10 bits give 0 to 1023. This number, coupled with a clock rate and pulse discharge rate suitable for the magnitude of incoming pulses (virtually all commercial amplifier systems produce output pulses in the 0–10 volt range), defines the size (number of channels) of the MCA, except in increasingly rare cases in which counts in only a portion of the total measured range are counted (as memory has become ever less expensive, larger MCA memories have become common). The energy span defined by the amplifier gain, the clock speed and number of bits (the conversion gain of the ADC), together define the spectrum calibration, usually expressed in electon volts per channel.

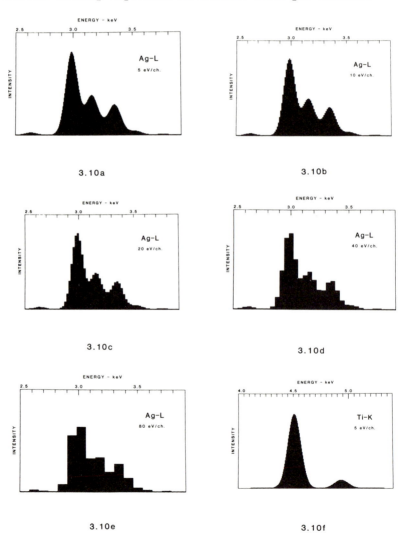

3.10a

3.10b

3.10c

3.10d

3.10e

3.10f

34

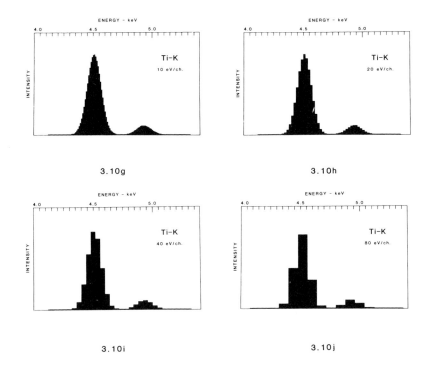

Figure 3.10 (a)-(j) The definition of peak shape as stored in (and displayed by) the MCA depends on the calibration, usually expressed in electron volts (eV.) per channel. For isolated peaks as in the Ti K line series, rather coarse definition may be acceptable. For overlapped peaks such as the Ag L lines, more channels are needed.

The calibration is important in that it allows us to convert peak position to energy for identification, and should be the same for spectra we may wish to compare. Also, there must be enough points (a smaller eV/channel value gives more points) across a given peak to properly describe its shape and allow us or the computer to find irregularities that may be another, partially hidden peak. Most systems are calibrated to specific integral values, such as 10, 20, 25 or 40 eV./channel. Figure 3.10 illustrates the appearance of simple and complex peaks at various calibration settings. The calibration process requires setting the amplifier gain and either amplifier zero (baseline voltage) or ADC zero (crossing point on rundown to stop the clock) so that peaks of known energy (from known elements) lie at positions where they belong. Since peak energies rarely are integrally divisible by the desired calibrations, they will usually not lie in exact channels, but rather between them. This means that locating the peak position well enough to assure proper calibration is difficult for many operators, and yet miscalibration that places a peak as little as 5 eV. (one quarter of a channel at 20 ev./channel) from its proper position will cause large errors in quantitative data reduction. Consequently, most systems use a computer-aided approach in which the operator designates peaks to be used for calibration, and enters their identity or energy, and the attached computer finds the

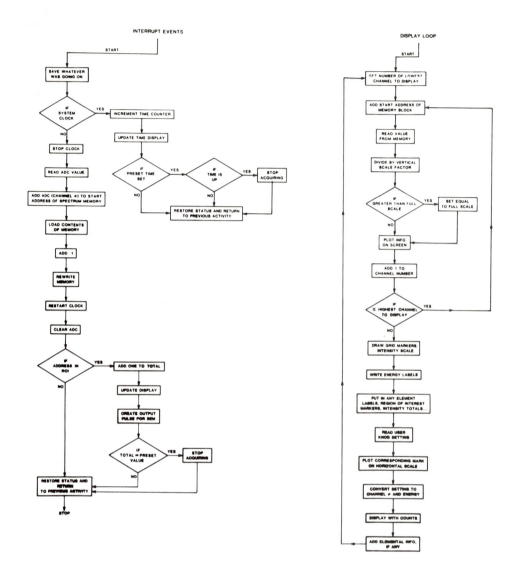

Figure 3.11 Simplified block diagram of a representative MCA program. The display loop runs all the time, generating the display of the spectrum with its various labels and markers. Any interactive response to user inputs are also handled there. The interrupt portion of the program is carried out for each pulse processed by the ADC, which signals the MCA computer when the data are ready.

actual centroid, usually by fitting a parabola to some number of channels. From the actual position of two peaks (assuming perfect linearity in the amplifier and ADC), the change in gain and zero needed to bring the system into proper calibration can be determined and the adjustments made either manually or automatically.

The MCA, whether it consists of a program running in a computer or hard-wired circuits, takes the number generated by the ADC, uses it as a memory address, and adds one to the number of counts stored at that address. Thus do the peaks grow with time. In a computer environment, the memory used may be part of the general purpose memory of the entire computer, or it may be a separate memory used only for data storage. The size of the memory (number of bits per channel) defines the maximum number of counts that can be stored. This may range from 16 bits (65535 counts per channel), adequate for many electron-excited applications where low count rates make it unlikely that this value would be exceeded, to 24 bits (over 16 million counts), adequate for virtually any circumstance involving even high count rates as in X-ray fluorescence with long counting times.

The MCA usually carries out other functions as well, including preset time termination of analysis. It may also integrate the number of counts in a particular energy range (called a window or region of interest, and usually covering a peak of an element or section of background) and (for SEM work in particular) produce an output pulse for each count stored in the region which is used to brighten a dot on the SEM display to "map" the location of the element on the scanned surface. This and related techniques will be covered in a later chapter. Most MCA's also devote a great deal of their circuitry to the display of the spectral data as and after it is acquired, with user aids in the form of energy and elemental markers, color coding of various information, comparison of several spectra, and so on. There is at a minimum some method to show the spectrum, with selectable (sometimes automatic) vertical scale settings, perhaps including a logarithmic scale to compress large and small peaks for visual examination at the same time, and horizontal expansion to permit examination of particular portions of the spectrum of interest. Generally systems that incorporate computer processing capability use the same display to show the results of that work, including both graphic displays of the processed spectrum and alphanumeric listings of calculated results. Recording of these on paper, film, or in other ways is obviously useful.

To illustrate a typical MCA operation, Figure 3.11 shows a very simplified logic flow diagram for typical data accumulation.

Qualitative Analysis with ED Systems

Many ED systems, especially those used with electron microscopes, are called upon most frequently to provide qualitative, rather than quantitative analysis (the latter subject will be covered extensively in subsequent chapters). Qualitative analysis, the identification of what elements are present in the sample, is at first glance a very simple task, yet has some complications that may arise to make it extremely difficult in particular situations.

For most cases, the process of identifying the elements present requires the user to recognize the peaks in the spectrum as corresponding to those emitted by a specific element. Most modern MCA's not only label the spectrum display directly in energy, but also have some capability to display the position of the principal lines of the elements, superimposed on the measured spectrum. Often the operator can scan through the entire periodic table while watching the display, and observe which set of displayed lines matches the peaks shown. The match should be both in position and also relative height, for those energies which correspond to peaks from several elements. These might be different lines within one shell, such as the K-β line of titanium at 4.93 keV. which is very close to the K-α line of vanadium at 4.95 keV., but can be distinguished from it because the titanium line would be accompanied by a much larger K-α peak at 4.51 keV. while the vanadium peak would instead have a much smaller K-β peak at 5.43 keV. In other cases, the confusion may be between peaks from different shells, such as the phosphorus K-α at 2.05 keV. and the zirconium L-α at 2.04 keV. Again, consideration of the pattern of accompanying lines for each element will identify which is present. Figure 3.12 shows the peaks from sulfur K, molybdenum L and lead M X-rays. While the principal line of each element is at nearly the same energy, the pattern of additional lines is quite different so that the total peak shape is distinguishable. In addition, for the Mo and Pb, higher energy lines from other shells might also be measured for confirmation.

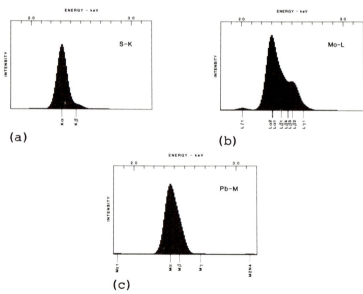

Figure 3.12 (a) Sulfur K, (b) molybdenum L and (c) lead M peaks.

Some systems will also directly search the tables to find all elemental lines that lie within a narrow energy band (to accomodate small errors in calibration) of a peak energy, and present the lines of each element in turn for the user to examine. This method is usually quite adequate for major element peaks. The user is then supposed to note the presence of the element, perhaps labelling the peaks on the display, and then disregard them, continuing on to the next element. Usually the procedure is carried out starting with the largest peak, and continuing to smaller ones. Eventually the point is reached where there are no more peaks clearly present above background, or there are some peaks that cannot be readily identified because they are minor lines which were either not included in the tables, or represent the various spectral artefacts described earlier. The trace element peaks can be improved somewhat by extending the analysis time, or perhaps by changing the conditions (eg. accelerating voltage), but these effects will be discussed later on.

Unfortunately, the simple sequence of logic just described is not always adequate. The breadth of the peaks in the energy dispersive spectrum is great enough that the probability of significant overlaps occuring is quite substantial. If overlaps occur between minor lines of major elements, they will probably be recognized without much difficulty. But overlaps between major and minor element peaks can completely hide the presence of the minor element. Figure 3.13 shows an example of such a problem. Titanium K and barium L lines are easily distinguished when present singly, because of the different pattern of $K\alpha - \beta$ and $L\alpha - \beta - \gamma$ peaks. In a sample with a 4:1 ratio of Ti to Ba, the barium peaks are still recognizeable, even though the largest peak is covered by the Ti K-α peak. It does require somewhat more judgement by the operator to recognize the presence of the barium, however. And the consequence of the overlap is that other lines from low concentration elements such as vanadium K or cerium L would be quite indistinguishable.

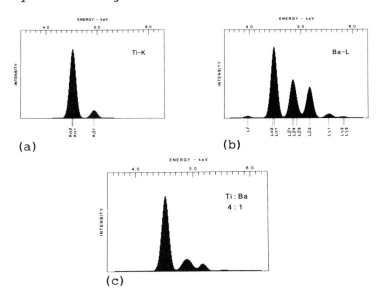

Figure 3.13 Overlap of Ba-L and Ti-K lines: a) Ti -K, b) Ba -L, c) 4:1 Ti-Ba

In the case of the sulfur – molybdenum – lead confusion shown before, if either Mo or Pb were present it would be very difficult to identify the presence of the sulfur. The comparison of the Ti K $-\alpha$ to K $-\beta$ just mentioned will always give a ratio of peak heights within a few percent of the nominal 8:1 ratio shown in most tables. This can be altered slightly by matrix absorption, but rarely enough to confuse identification. But for the lead or molybdenum, it is not possible to confidently compare the height of the low energy M or L peaks to the height of higher energy peaks from a different shell (the ratio depends strongly on the choice of analysis conditions and on the matrix composition). Consequently, it is not practical to judge from the height of the peak whether or not a small amount of sulfur is present and is making the observed Pb or Mo peak larger than it should be. Separation of such overlaps, even if they are recognized as being present, requires the same approach as will be described for quantitative analysis.

The key, of course, is recognition. If the user expects to find sulfur along with molybdenum (for instance in a steel inclusion) he may recognize the peak at 2.3 keV as being a mixture of both Mo and S, and proceed to carry out the necessary peak deconvolution. But if the element is not anticipated, it may be missed altogether. Figure 3.14 shows a typical stainless steel spectrum. The major elements are Cr, Fe and Ni. The K-β peak from the chromium is much larger than, and competely hides, the small peak from the 1% or so manganese likely to be present. The even smaller Mn K-β is competely covered by the Fe K-α . Unless the analyst expects to find manganese, it will be missed. Even in a quantitative analysis, the presence of the manganese might be overlooked without its abscence being obvious in the final results.

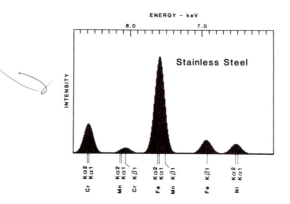

Figure 3.14 Stainless steel spectrum in which the Mn peaks are hidden.

This also illustrates that detection limits with ED spectrometers must be taken in context of the sample matrix. It is necessary to detect a peak for the element not only above background, which will vary substantially depending on how the sample is excited (with charged particles such as electrons or with X-ray or γ- ray photons), but also above any other peaks that may be present. With X-ray excitation, in a matrix of silica (a typical mineral, perhaps), manganese detection limits of a few parts per million are readily obtained. With the same excitation, in a stainless steel, we have just seen that 1% can be missed. Electron excitation generaly gives poorer detection limits because of the higher background levels, but this is unimportant if the ultimate limit is set by interfering peaks from other elements.

There has long been interest in methods that would provide automatic peak identification (Moeller,1982), in order to report a list of detected elements without operator intervention. These methods typically rely on extensive computer processing to carry out a cross-correlation of the spectrum with a typical or idealized peak shape. Cross-correlation (Connelly, 1972) is a readily computerized procedure that multiplies (channel by channel) a portion of the spectrum by a list of values that represent a peak shape (sometimes simplified drastically to a simple rectangle). The sum of the products will be quite large if a peak is really present, and much smaller if it is not. Performing this operation at each channel position through the spectrum will identify the presence of significant peaks more positively than simply looking for values in the spectrum itself that exceed a fixed threshold. This locates peaks and delivers a list of energies. Alternate methods that use first, second, or even higher order derivatives of the spectrum to locate peaks work well for major peaks, but often have difficulty with small peaks which have significant irregularities due to counting statistics and are hard to distinguish from background (with its own counting statistics) (Barbi, 1980).

In any case, the list of energies is still a long way from the desired list of elements. An extensive logic tree to test for the proper presence of minor lines for various possible elements will identify most of the major elements with little difficulty. For the remaining confused or overlapped peaks, such as those shown in the examples above, or peaks from low concentration elements, the automatic methods are not yet reliable. Development of more extensive computer models for fitting of spectra, as used for deconvolution, and better ability to predict ratios of peaks, especially between different shells, will certainly help, and the availability of more and more computing power will permit the implementation of these methods. Nevertheless, at least for the present, the most reliable qualitative analysis computer to recognize peaks in spectra is an operator who understands the instrument and its characteristics, and has some independent knowledge about the sample and what elements may reasonably be expected to be present in it.

Chapter 4

Electron Penetration in Solids

When a beam of electrons strikes a bulk sample, as in the SEM or electron microprobe, the individual particles follow different paths, depending on the random nature of their encounters with atoms in the material. It is all too easy to visualize this as though a steady beam of electrons was shining on the sample, and a cloud of them was busy moving about in the solid. This is fortunately far from the true situation, since the electrostatic repulsion between the electrons would considerably complicate our job of mathematical description. In fact, the SEM in particular has such low beam currents that most of the time there are no electrons in the machine at all (except in the immediate vicinity of the electron gun), and from time to time a single electron travels down the column, strikes the sample, causes whatever signals we observe or other processes to take place, and then quiet reigns again. There are higher beam currents present (more electrons) in the TEM and microprobe, but even there we can safely consider the electrons on a one-at-a-time basis. At typical beam accelerating voltages, the electron velocity is a sizeable fraction of the speed of light, so the travel time along a metre or less of column is quite short.

The electron moving in a solid can lose energy in several ways, the only ones important for our immediate concerns being the generation of Bremsstrahlung by interactions with the nuclear field, and the ionizations of inner shells of the electrons around the atoms, which may give rise to characteristic X-ray emission. It is these mechanisms for energy loss that determine the range of the electrons in the material. Collectively they are called the stopping power 'S', and obey the Bethe (1930) relationship:

$$S = -7.85 \ 10^4 \ (Z/A) \ (1/E) \ \log (1.166 \ E \ / \ J \ )$$

giving S in units of keV./cm. where E is the electron energy and Z is the atomic number, A the atomic weight and J the mean ionization potential of the target element. The mean ionization potential represents an average over several shells, and has been approximated by several expressions:

1) (Bloch, Wilson 1941)
   $$(J/Z) = 11.5 \ eV.$$

2) (Berger & Seltzer, 1964)
   $$(J/Z) = 9.76 + 58.5 \ Z^{-1.19} \ \text{(for } Z>12)$$

3) (Duncumb et al., 1969)
   $$(J/Z) = 14 \ (1-e^{-0.1 \ Z}) + 75.5/Z^{Z/7.5} \ -Z/(100+Z)$$

These are summarized in Figure 4.1. The Duncumb & Da Casa expression
is perhaps a classic example of something we shall encounter
frequently in this field: expressions that have been fitted
mathematically to measured data but whose functional form is derived
from mathematical fitting considerations or the intuition of the
researcher largely without regard to any possible underlying physical
significance. In this particular case, the purpose of the expression
for J was to force better agreement between computed quantitative
analytical results and known standard compositions, particularly for
light elements. It is far from clear that it was the J values that
were in error and caused this bias in the quantitative results, but by
using the expression shown, a better overall average accuracy was
obtained. From the figure, it can be seen that this expression changes
dramatically at low Z. Quasi-theoretical arguments have (post facto)
been brought forward to explain why the mean ionization potential
might behave in this way, as a consequence of filling of shells and
shielding effects in the electron cloud. It is probably better to
accept the equation for what it really is: a useful way to compensate
for things we do not know in detail or model in our equations, which
allows us to get acceptable results. Both the Bloch and the Berger &
Seltzer expressions behave in more understandable ways, and it is the
latter that is most used in present day correction programs.

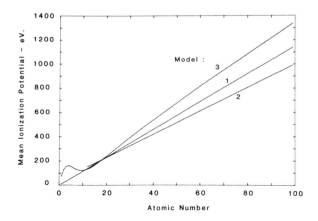

Figure 4.1 Mean ionization potential J as a function of atomic number,
comparing various mathematical models.

The stopping power S is a function of energy, as well as target
element. For a given element, as shown in Figure 4.2, the rate of
energy loss is low at first, and then increases as the electron slows
down (this is the same behavior as paying off the principal on a
mortgage).

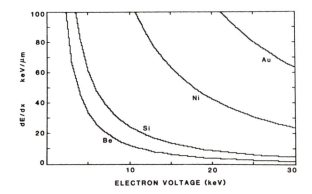

Figure 4.2 Rate of electron energy loss according to the Bethe law, as a function of electron energy and matrix element.

Since the electron energy decreases during the entire process of stopping, the range is given by an integral, actually:

$$R = \int_0^{E_0} (dE/\rho S)$$

where ρ is the density of the material (a constant which is convenient to pull out of many expressions). This integral can be closely approximated by:

$$\rho R = k \, E_0^n$$

where k and n are nearly constant parameters. Since we are interested only in the depth to which the electron penetrates until its energy drops below E_c, the critical excitation energy for the element and shell of interest, we may further stretch the approximation to:

$$\rho R = k \, (E_0^n - E_c^n)$$

values for the constants have been proposed by several researchers. For ρ expressed as grams/cubic centimeter, for E in keV., and for R in micrometers, we can choose between:

1) (Anderson, 1966)   k = 0.064   n = 1.68
2) (Reed, 1966)   k = 0.077   n = 1.5
3) (Castaing, 1960)   k = 0.033 A/Z   n = 1.5
4) (Kanaya, 1972)   k = $0.0276 A/Z^{0.889}$ n = 1.67

The differences are not great, and reflect the different voltages, elements and methods used by each person. Figures 4.3 and 4.4 compare them for the case of excitation of iron K α X-rays in iron, as a function of accelerating voltage, and for various elemental K lines (in the pure element) at 15 kV. accelerating voltage.

44

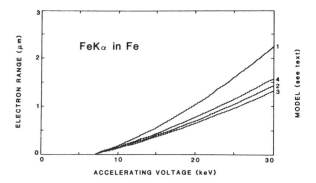

Figure 4.3 Electron range for generation of Fe K-α X-rays in iron as a function of accelerating voltage, by various models.

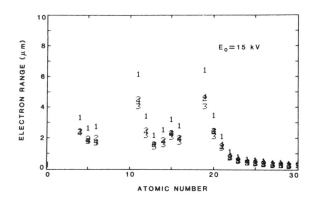

Figure 4.4 Electron range for generation of K-α X-rays in pure element bulk targets, for 15 keV accelerating voltage, by various models.

Reed has in addition created a nomogram as a graphical aid in applying his expression, as shown in Figure 4.5. To use it, begin with the critical excitation energy for the element of interest (the example shows the case for iron, about 7 keV.). Then proceed to the accelerating voltage being used (20 keV. in the example), through the density (iron metal is about 7 g/cm^3) to the approximate depth of analysis (0.8 micrometers). This is of course only approximate, and does not tell us anything about the shape of the excited volume or the distribution of X-ray generation with depth, but it can give a good insight into the scale of our 'microanalysis' and indicate whether the features of interest are large enough to contain the excited volume.

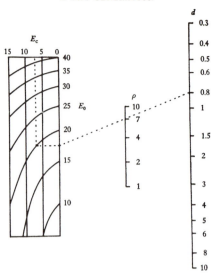

Figure 4.5 Nomogram of Reed to estimate electron range in bulk samples, as a function of accelerating voltage E_o, elemental absorption edge energy E_c, and sample density ρ (Reed, 1966).

The electrons do not travel in a straight path while slowing down. All of the important energy loss mechanisms involve some change in direction. The Bremsstrahlung events are also called large-angle scattering, since the electron may be deflected considerably in passing through the electrostatic field around the nucleus. Ionization of electron shells causes small-angle deflections in the path. In mathematical modelling of the electron trajectory, to be discussed in more detail shortly, the electron path is considered to be a series of straight line segments between large angle events, with the stopping power (which describes the small-angle ionization events) acting to slow the electron down only along the straight segments. In other words, although the large angle scattering events are responsible for all of the spectrum background, and for the direction the electrons take (which controls the shape of the excitation volume), it is ignored as far as energy loss of the electrons is concerned because it accounts for so little of the total original energy of the electron.

The size of the excited volume (the electron range) depends on voltage, as we have seen. The shape depends on large angle scattering events, which in turn depend on the atomic number of the target element (or the average for real samples which consist of several elements). For low Z materials, the depth of the excited region is greater than the width, and conversely for high atomic number materials the excited volume is reduced in depth and greater in width, as shown in Figure 4.6. In most cases, it is adequate to picture the excited region in which the incident electrons do their work as approximately spherical, with something of a water-drop neck forming for lighter elements, and some squashing down for heavier ones. If you

46

remember that this shape is dependent on atomic number, while size is independently a function of electron accelerating voltage, you will have the conceptual basis for understanding much of what happens in electron-excited X-ray analysis.

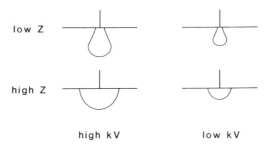

low Z

high Z

high kV   low kV

Figure 4.6 Diagram of electron capture volume in bulk targets. The shape varies with atomic number and the size with accelerating voltage.

For different elements in a multi-element sample, the depth of excitation for each will be different, and less than the total depth of electron penetration. Figure 4.7 illustrates this for the case of a sample of copper and aluminum. The copper K excitation edge is at 8.98 keV., so any electron which drops below that energy can no longer ionize the copper K shell. Enough energy may well remain, however, to excite the aluminum K shell, which requires only 1.56 keV. Electrons will, on the average, be able to penetrate deeper into the specimen before their energy falls below the Al critical energy than below the Cu energy, so the analysis of aluminum covers a larger and deeper region. This can be important in dealing with materials which may be inhomogeneous in depth, in which one element or another may be excited only near the surface, or vice versa.

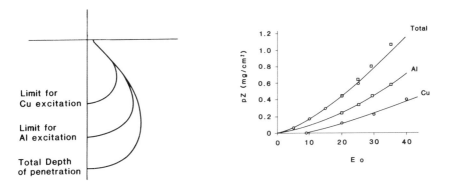

Figure 4.7 Depth of analysis for Cu and Al in a composite sample (Shinoda, 1968).

47

Knowing the depth and shape of the excited region still does not define where the emitted X-rays come from. The probability of ionization of an atom by an electron depends on the electron's energy, which is changing. Generally, for a given element the probability, called the ionization cross section, increases as electron energy drops (until it falls below the critical energy, of course). Combining this information with the probability of the electron turning from its original direction, it is possible to sum up the ionizations that occur at each depth in the sample. A plot of excitation as a function of depth is called a ϕ (ρ z) curve. and has the general appearance of the one shown in Figure 4.8. The curve initially rises with depth below the surface since the electrons spend little time there (they are going fast and heading down) and they have a low cross section because their energy is still high. The maximum excitation is produced at some depth below the surface, and then the curve decreases as fewer and fewer electrons reach to great depths (because large angle scattering has redirected them).

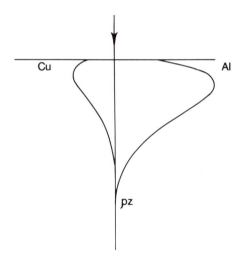

Figure 4.8 Depth distribution of excitation ϕ (ρ z) for 90% Cu, 10% Al alloy with a 20 keV electron beam.

The shape of ϕ (ρ z) curves have been measured for a number of elements using thin layers of a tracer element under varying thicknesses of another element. From many of these curves, as well as from simulations carried out with 'Monte-Carlo' programs in computers, a general picture of the shape of these curves has emerged, and has been fit to a non-theoretical but very useful model. In terms of the depth z and the density ρ, Brown and his co-workers (1979a) have reported the expression (Parobek, 1978):

$$\phi(R) = DKn \ (KR)^{n-1} \ exp(-(KR)^n)$$

where $R = \rho z + \rho z_0$. The parameters D, K, n and ρz_0 are calculated as functions of atomic number Z, atomic weight A, accelerating voltage

E_o and critical excitation energy E_c. The functions are not theoretically based, but were obtained by application of a Simplex optimization program, a mathematical technique which minimizes the differences between computed values and experimental data when the functional dependence is non-linear. Their results are:

$$n = 1.95 \text{ A} / z^{1.32}$$

$$K = B \text{ F} / (E_o - E_c)^m$$

where

$$B = 0.5409 / E_c^{1.228}$$

$$F = 1 + (0.05199 \ z^{0.7} / \exp (0.0002 \ z^2))$$

$$m = (15.6578 / E_c)^{0.2505}$$

$$D = Q (E_o - E_c)^P$$

where
$$Q = 0.5945 \ E_c^{1.42} \ z^{0.26}$$

$$P = 2.111 \ (1.84/E_c)^{0.325}$$

$$\rho z_o = 24.487 \ E_o \ E_c^{0.428} / z^{1.259} \ (1+(1/\exp(0.00015 \ Z)))$$

From this equation it is straightforward to predict the depth distribution of excitation in samples of known composition, at least for the cases of perpendicular electron beam incidence. Knowing the depth, we can make absorption corrections (as will be discussed shortly) or simply confirm that the analyzed volume is within a feature of interest. The curves show that different elements in the same sample are analyzed at different depths, as was illustrated in the previous figure.

Electron Backscattering

The redirection of electrons by large angle scattering in the field near the atomic nucleus can turn them around so that electrons which initially entered the sample perpendicular to the surface may actually leave the surface again and be lost to the process of ionization. Since the probability and angle of large angle scattering both increase with atomic number of the target element, the total amount of backscattering increases with atomic number. Figure 4.9 shows a plot of η (the fraction of electrons that backscatter) as a function of Z. This is a somewhat idealized curve, with some elements showing minor scatter about the smooth curve because of the effect of A/Z (atomic weight/number) on stopping power. It can be estimated rather well as:

1) $\eta = \log (Z) / 6 - 1 / 4$

and is quite independent of electron accelerating voltage in the range of voltages used in electron microscopes (because the scattering angles and probabilities are independent of keV. in this range and depend only on Z).

Somewhat more complex expressions due to Reuter (1972)

2) $\eta = -0.0254 + 0.016Z - 1.86 \times 10^{-4} \, Z^2 + 8.3 \times 10^{-7} Z^3$

and Heinrich (1981)

3) $\eta = 0.5 - 0.228 \times 10^{-4} \, (80-Z) \, (ABS(80-Z))^{1.3}$

are compared to the simple expression in Figure 4.9, along with some measured values.

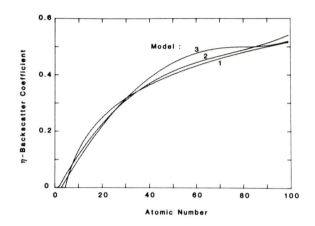

Figure 4.9 Fraction of incident electrons which backscatter, as a function of atomic number, comparing various models.

Robinson (1982) has proposed that these values can be adjusted for non-perpendicular incidence of the electron beam using the following function:

$$\eta(\phi) = 0.891 \, (\eta \, (perp)/0.891)^{\sin(\phi)}$$

in which the angle ϕ between the incident electron beam and the surface is 90 degrees for perpendicular incidence. It seems likely that there is some interaction between inclination and atomic number, however, owing to the effect of atomic number on the shape of the excited region.

In order to correct the prediction of excited X-ray intensity for the backscattering of electrons, it is not enough to know how many of them backscatter. We also need to consider how much energy they carry away with them (after all, if the backscattered electron had already lost energy during its passage through the sample, and dropped below the excitation energy of the element of interest, we would not care if it backscattered). The electrons that do backscatter from the sample do not, on the average, penetrate as deeply as those that do not. Again, this depends on atomic number, but generally the range of

dependence of R on Z and U. Note that as we would expect on the basis of qualitative arguments about electron scattering and the role of Z and energy, that R decreases as Z increases (more backscattering) and as U increases (the backscattered electron is more likely to carry away useful energy, being still above the element's excitation energy).

Rather than use the graphical data determined by experiment, we require an equation with which to calculate R from Z and U, in order to perform quantitative corrections. Several such expressions exist, as shown below. None has any physical basis for the functional form, but rather depend on the intuition of the person doing the fitting and on his access to computer facilities to carry out the regression.

1) Duncumb & Reed (1968)

$$R = 1 + 10^{-2} Z (-0.581 + 2.162W - 5.137W^2 + 9.213W^3 - 8.619W^4 + 2.962W^5)$$

$$+ 10^{-4} Z^2 (-1.609 - 8.298W + 28.79W^2 - 47.74W^3 + 46.54W^4 - 17.68W^5)$$

$$+ 10^{-6} Z^3 (+5.400 + 19.18W - 75.73W^2 + 120.1W^3 - 110.7W^4 + 41.79W^5)$$

$$+ 10^{-8} Z^4 (-5.725 - 21.65W + 88.13W^2 - 136.1W^3 + 117.8W^4 - 42.45W^5)$$

$$+ 10^{-10} Z^5 (+2.095 + 8.947W - 36.51W^2 + 55.69W^3 - 46.08W^4 + 15.85W^5)$$

where $W = 1/U$

2) Yakowitz et al. (1973)

$$R = A - B \log (CZ + 25)$$

$$A = 0.00873U^3 - 0.1669U^2 + 0.9662U + 0.4523$$

$$B = 0.002703U^3 - 0.05182U^2 + 0.302U - 0.1836$$

$$C = 0.887 - 3.44/U + 9.33/U^2 - 6.43/U^3$$

for $U \leq 10$

3) (Russ)

$$R = e^{-0.004 Z (U-1)^{0.3}}$$

Figures 4.13 and 4.14 illustrate these models for several cases. Note that the agreement is quite good for moderate overvoltage ratios, but is poorer for very high values of U. It also becomes poorer for high atomic numbers.

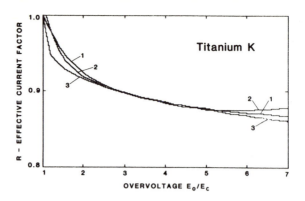

Figure 4.13 Effective current factor R as a function of overvoltage for titanium.

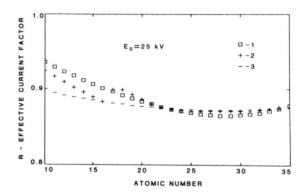

Figure 4.14 Effective current factor R as a function of atomic number for excitation of K-shell with 25 keV electron accelerating voltage.

For non-normal incidence of the electron beam, it is clear that R will decrease because both the number and energy of the backscattered electrons increases. It is very tempting to assume that the functional dependence on the tilt angle might be separable from the dependence on Z and U, or in other words that R(tilt) = R(normal) x function(tilt). Since the shape of the $d\eta/dE$ curve changes with tilt, this is not really true. However, there is not enough experimental data available to permit fitting of universal functions to Z, U and S (surface tilt angle, zero for normal incidence). In practice, therefore, it has become common to either ignore the tilt effect (hoping it will cancel out between standards and unknowns if they are at the same angle), use the function shown below to modify the calculated R for perpendicular incidence, or if necessary resort to complete Monte-Carlo modelling to get R for the tilted case, although this is impractical for unknown samples. The equation most often used was obtained by Heinrich (Fiori,

54

1976a), fitting to experimental data taken at 45 degrees. It does include a dependence on U.

$$R = R(perp) \exp(\log(0.0753/U+0.8994) \ (\sec(S)-1.0+S/90))$$

Another line of reasoning can also be used, based on our previously introduced simple model for electron penetration, the shape of the excited volume, and stopping power. Let us assume for the moment that all of the incident electrons initially penetrate to some depth X with no energy loss, and then radiate outwards uniformly in all directions, producing X-rays as they go, with a constant stopping power. It seems at first glance as though this model is completely inadequate to describe the details of what we know to be a very complex set of interactions. However, the model was actually proposed at one time as suitable for calculating a depth distribution of X-ray production $\phi(\rho z)$ adequate for making absorption corrections. In any case, this model may still be useful to understand the trends and gross effects of surface inclination on R.

The dummy variable X can be related to the effective current factor R if the stopping power (rate of energy loss with distance) of the electrons is independent of energy. Then the portion of the sphere which protrudes from the surface represents the fraction of energy that escapes through backscattering, as shown in the sketch in Figure 4.15, and by integration we can get R in terms of X. The total energy in the sphere is:

$$\int_0^1 4\pi r^2 \ (1-r) \ dr = \pi/3$$

while the energy in the protruding segment is:

$$\int_X^1 2\pi r \ (r-X) \ (1-r) \ dr = \pi(1/6 - X/3 + X^3/3 - X^4/6)$$

which gives for R (which is 1 minus the energy in the segment divided by the total energy):

$$R(perp) = 1/2 + X - X^3 + X^4/2$$

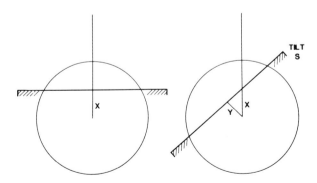

Figure 4.15 Diagram of model to predict the effect of surface tilt on effective current factor.

This relates the effective current factor to the dummy variable X. Now if the surface is inclined by an angle S, as shown in the sketch, the original electron penetration distance X is unchanged, but more of the sphere protrudes from the surface, and we shall interpret this as giving the increased loss of energy (reduction in R) with tilt. The distance Y is just X cos (S), and R becomes:

$$R(\text{tilt}) = 1/2 + Y - Y^3 + Y^4/2$$

Eliminating X and Y gives a set of relationships between R for perpendicular incidence and R for a surface tilted by an angle S, as shown in the plot below (the superimposed data were measured at 45 degree incidence). This model gives complete separation between the dependence of R on Z and U and the effect of tilt. It is clearly oversimplified, yet offers reasonable agreement with measured data and is simple enough for incorporation in any calculation model involving surface tilts. More than that, it may offer another conceptual hook upon which to hang our mental picture of the electron/sample/X-ray interaction process.

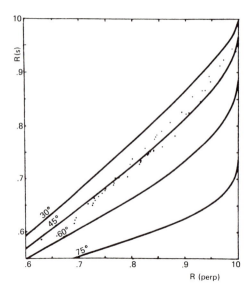

Figure 4.16 Calculated effective current factor R for surface at various surface tilt angles, as a function of the R factor for normal incidence, compared to measured data for 45 degrees (Russ, 1975).

56

Chapter 5

Monte-Carlo Modelling

In the preceding chapter a model was described to predict the depth of penetration, backscattering, and energy loss of electrons and the subsequent distribution of X-ray generation. It applies only to the case of semi-infinite (infinity comes soon for electrons, within a few micrometers) homogeneous bulk solid with a flat surface and a perpendicularly incident electron beam. It happens that while such specimens are not unknown, they represent only a fraction of the interesting kinds of things we would like to analyze. Since the equations combine many discrete effects and attempt to describe average behavior, it is impossible to introduce modifications that will enable them to adapt to the infinite variety of boundary conditions encountered in the "real world." We shall later on see that useful quantitative models can be extracted for a few situations, such as thin sections and (with less accuracy) coatings or particles on a substrate. For more general cases we need a different approach altogether.

The technique that is available (and used in many other unrelated modelling problems) is Monte-Carlo. This method, named after the famous gambling resort because it depends on the generation of random numbers and 'luck' to predict what happens (and we know that on the average people lose money gambling, but we can't predict from this the fate of any individual gambler). The Monte-Carlo simulation method (Heinrich et al.,1976) follows a large number of individual electrons (in this case) as they enter the specimen and undergo the various processes which can take place. At each point in their history where a choice or series of choices must be made, a random number generated by a computer (which is keeping track of everything that happens - the electron is just a figment of its imagination) is used to select from a probability distribution function that defines the various possible outcomes and their probabilities.

For an electron, the most important events are encounters with the atoms in the material. The kinds of scattering (elastic and inelastic) and the physical description of these events including the amount of energy loss and the angle of scatter can be accurately modelled by simple equations. For each electron and each encounter, a slightly different outcome is predicted using random numbers. From many such encounters the bulk averages will emerge. But in the complete Monte-Carlo simulation the entire trajectory of each electron, one at a time, is worked out step by step. From each large angle scattering event, the new direction is calculated. These determine where the particular electron goes, and it is easy for the program to tell if it backscatters from the sample (and what its direction and energy were when it did). Energy loss along the path the electron follows is computed using the Bethe relationship seen before. This means that the energy of the electron is known at each position, and from that the cross section for inner shell ionization to produce X-ray emission. The X-rays, once 'generated' can be treated by conventional (analytical) equations for absorption and so forth, as will be described in the next chapters.

Sums can easily be built up showing the distribution of electron energies, ionizations, or anything else of interest as a function of depth in the specimen, angle above it, and so forth. Specific Monte-Carlo programs have been used to compute the energy and angular distribution of backscattered electrons, the production of secondary electrons and the effect of magnetic contrast, and so on. These results can be compared to measured data and to the distributions predicted by the analytical expressions, which serves as a good control on our entire mathematical description of these processes. The importance of the Monte-Carlo technique in assisting in the refinement of conventional equations would by itself justify its inclusion in this book.

Of even greater importance is the ease with which Monte-Carlo simulation accommodates non-ideal boundary conditions. If the sample surface is inclined to the beam, or is not flat, or of the sample is a thin section or a coating on a substrate, or even if it is a mass of separated fibers, clumps of tiny particles of different composition, or worse, Monte-Carlo can handle it. As each electron trajectory is followed, step by step, the program always knows where it is, in what part of the sample it is (hence the composition and from that the various probabilities of scattering, energy loss, ionization, etc.), and when it leaves the sample through any side, top, bottom, and so on.

Figure 5.1 shows the typical calculations that are performed. The velocity and direction of the particle, as well as its present coordinates, are calculated after each large angle scattering event. These are all considered, in most programs, to be elastic. In other words, the production of continuum (Bremsstrahlung) is ignored unless it is the Bremsstrahlung production that is itself of interest. This is justified in practically all cases encountered in the electon microscope because although the spectrum background is very significant in our efforts to find small peaks, its production accounts for only a tiny fraction of the the total energy deposited in the sample by the electrons, and ignoring it does not alter the simulation of characteristic X-ray production by ionization. The large angle scattering events make the simulated electron trajectory into a three-dimensional zig-zag. Along the straight segments of this path, the electron's energy loss is modelled in terms of the matrix stopping power (a function of the electron's remaining voltage).

The main procedures in Monte-Carlo simulations are thus to calculate the new direction and step length after each large angle scatter and the energy loss and probability of ionization along each step. For the direction calculation, the angles θ and ϕ are easily obtained from a computer-generated random number (between 0 and 1). The ϕ angle has a constant probability distribution, that is any value is equally likely (notice that transforming the coordinate

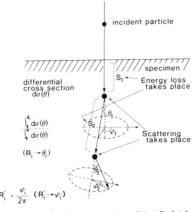

incident particle

specimen

S_1 ← Energy loss takes place

differential cross section $d\sigma(\theta)$

$\dfrac{\int_\theta^\theta d\sigma(\theta)}{\int_0^\pi d\sigma(\theta)}$

$(R_i \rightarrow \theta_i)$

Scattering takes place

$R_i' = \dfrac{\varphi_i}{2\pi}$ $(R_i' \rightarrow \varphi_i)$

R_i, R_i' : Random Number $(0 < R \leq 1)$

ELECTRON

Elastic Scattering

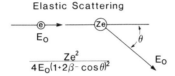

E_0 (Ze) θ

$\dfrac{Ze^2}{4E_0(1+2\beta - \cos\theta)^2}$ E_0

Inelastic Scattering
(Continuous slowing down approximation)

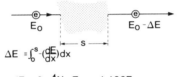

E_0 $E_0 - \Delta E$

$\Delta E = \int_0^s -\left(\dfrac{dE}{dx}\right)dx$

$\dfrac{dE}{dx} = \dfrac{2\pi e^4 Na}{E}\dfrac{Z}{A}\ln\dfrac{1.166E}{J}$

Figure 5.1 The Monte-Carlo modelling process follows individual electrons along paths whose scattering angles are estimated probabilistically, while computing the rate of energy loss due to inelastic scattering and characterstic X-ray production (Shimizu, 1977).

system after each collision to keep our polar coordinate system with us simplifies these calculations enormously). The θ (deflection) angle can either be calculated or sampled from a tabulated probability distribution, based on the generated random number. The simplest expression for the scattering angle is the Rutherford cross-section, which gives the probability of scattering through an angle greater than θ as

$1.62 \ 10^{-20} \ Z^2/E^2 \cot^2 (\theta/2)$

events/ electron-atom-cm.^{2}. It can also be written in terms of the differential probability of scattering at the angle θ , based on atomic number Z and electron energy E and a screening factor a, as

$$d\sigma = Z \ e^2/4 \ E \ (1+2a-\cos\theta)$$

The screening factor a is, like the mean ionization potential we encountered before, a convenient place to hide uncertainty (or put more delicately, it makes a good adjustable parameter). A typical expression for the screening factor is:

$$3.4 \ Z^{2/3}/E$$

From this cross section, the angular scattering distribution as a function of angle can be determined. The detailed equations are not necessary for our conceptual understanding and are not included here. In the simplest form, the angular deflection is determined from:

$$\cot(\theta/2) = 2 \ R^{1/2}\alpha/ \ b$$

where R is a random number (0<=R<1), b = .0144 Z/E (for E in keV.) and α is an impact parameter chosen to make the backscatter factor computed for the ideal case agree with the experimental values. This is simple enough to permit the entire simulation to be carried out in very small computers, if you can wait long enough for enough trajectories to be accumulated to give averageable results. Figure 5.2 shows a representative flow chart for such a program.

This rather simple model with the approximate Rutherford cross-section, the continuous slowing-down approximation, and frequently a discrete number of steps covering the entire energy loss process of the electron, with average energies for each step used to compute step length and interaction probabilities, gives surprisingly good results for those cases (such as bulk samples) in which it can be compared to experiment (Curgenven, 1971; Love, 1977).

More elaborate simulations in which each single scattering event is accounted for using the more exact Mott cross-section, and the energy loss for each step is not taken directly from the Bethe law but is sampled from a probability distribution to model statistical fluctuations in the processes, are also being used. They are strikingly better in dealing with thin sections, where at most a few interactions of the electrons may take place. The calculations are only slightly more complicated, but so many must be performed for each electron that somewhat larger and faster computers are generally used (Heinrich, 1976).

Typical Results

Using even the simplest Monte-Carlo program, it is easy to demonstrate the effect of accelerating voltage on both electron range and the size of the excited volume (and generated intensity) for a given element, the effect of atomic number on shape of the excited volume, and the effect of beam incidence angle on the depth and shape

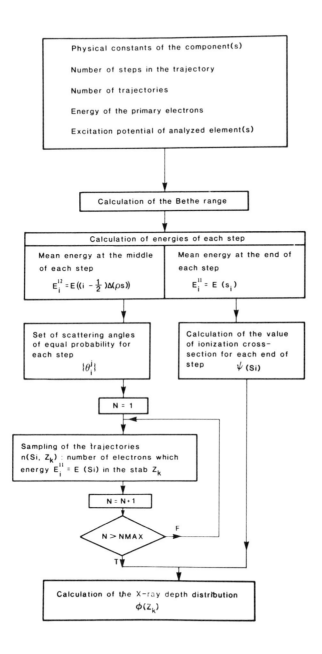

Figure 5.2 Flow chart for Monte–Carlo calculation.

of the excited volume. Examples of these are shown in Figures 5.3-5.5. Note that the electron trajectories are shown as zig-zag paths that become increaingly 'crinkled-up' toward the end. This is in agreement with our understanding of the behavior of the stopping power, which increases as electron energy decreases.

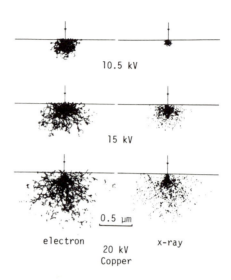

Figure 5.3 Monte-Carlo plots of electon trajectories and X-ray generation in copper at varying accelerating voltages (Newbury, 1976).

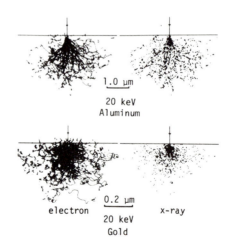

Figure 5.4 Monte-Carlo plots of electron trajectories and X-ray generation at 20 keV. in different target elements (Newbury, 1976).

62

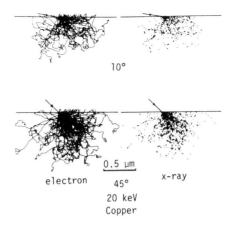

10°

0.5 μm

electron   45°   x-ray

20 keV
Copper

Figure 5.5 Monte-Carlo plots of electron trajectories and X-ray generation in copper at 20 keV. with different beam incidence angles (Newbury, 1976).

Analysis of particles and thin sections are becoming increasingly important, and can also be handled nicely by most Monte-Carlo programs. The effect of voltage on particles of various sizes, the increase of lateral size of the analysis region with thickness of a thin foil, and the effect of substrate atomic number on generation of X-rays from thin coatings are all indicated in Figures 5.6-5.8.

5 keV   17 keV   30 keV

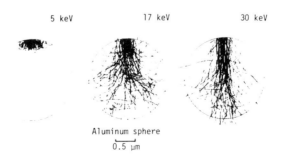

Aluminum sphere
0.5 μm

Figure 5.6 Monte-Carlo plots of electron trajectories in a spherical particle at different accelerating voltages (Newbury, 1976).

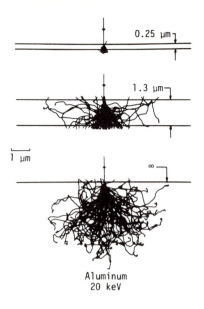

Figure 5.7 Monte-Carlo plots of electron trajectories in unsupported thin films of aluminum, of varying thickness (Newbury, 1976).

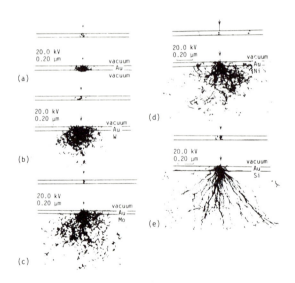

Figure 5.8 Monte-Carlo plots of electron trajectories in gold coatings on various substrates (Newbury, 1976).

In addition to these, most Monte-Carlo programs will sum the excitation as a function of depth to produce a ϕ (ρ z) curve. The effect of surface tilt angle on the shape of this curve is shown in Figure 5.9.

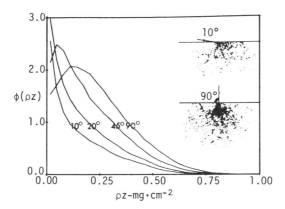

Figure 5.9 Depth distribution of X-ray generation in copper with 20 keV electrons, for varying angles of electron beam incidence (Newbury, 1976).

Combined with a discrete absorption correction for X-rays from each depth, using the expressions to be presented in the next chapter, these give depth distribution functions for both generated and emitted X-rays in various samples. Figure 5.10 shows that in a titanium-aluminum oxide, the aluminum atoms are excited to a greater depth than the titanium, as we would expect, but that because of matrix absorption, fewer of those generated at depth can be escape to be detected so that the actual analysis of the aluminum is carried out to a shallower depth than the titanium.

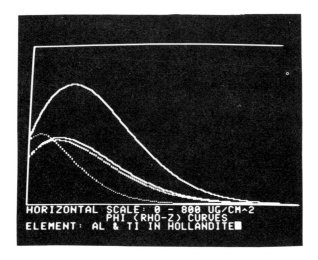

Figure 5.10 $\phi(\rho z)$ curves for Al and Ti in a ceramic. Solid lines show depth distribution of X-ray generation, and dashed line shows depth from which emitted X-rays come taking matrix absorption into account.

Finally, it is important to reiterate that it is in calculating electron energy distributions (backscattered, transmitted, secondary), and other distributions such as ϕ (ρ z), the F(x) function we shall encounter in the next chapter, and so on that fundamental knowledge is gained. Refinement of ionization cross sections, dependence of the backscattering coefficient on surface tilt, depth profiles of Bremsstrahlung production, and even the mean ionization potential have all been accomplished through comparison of experimental data with predictions from these simulations. Figures 5.11-5.14 show a few of these results.

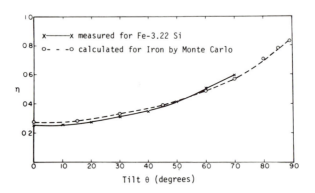

Figure 5.11 Backscatter coefficient as a function of specimen tilt, 30 keV electrons (Myklebust, 1976).

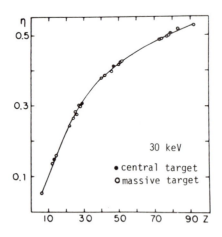

Figure 5.12 Backscatter coefficient as a function of atomic number, for 30 keV. electrons (Myklebust, 1976).

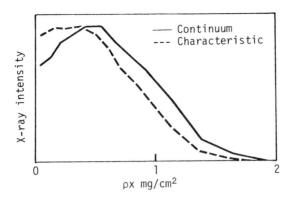

Figure 5.13 Distribution of 8 keV. continuum and characteristic X-ray production as a function of depth in a copper target, 29 keV electrons (Myklebust, 1976).

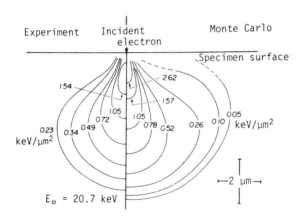

Figure 5.14 Energy dissipation profiles in polymethylmethacrylate, comparing Monte-Carlo calculations with experiment (Myklebust, 1976).

The Problem with Monte-Carlo

If Monte-Carlo simulation is so flexible and powerful, why don't we use it all the time? Admittedly the technique places more burden on the computer, and with a small mini-or microcomputer it can take quite a while to get enough trajectories for meaningful results with good statistics, but computers keep dropping in price and that does not seem like a very good reason to pass up such a powerful tool.

The problem is that Monte-Carlo simulation works the wrong way. It allows us to specify a sample geometry and composition, and then predicts for us what the measured X-ray and electron signals should be. In real life, we usually can measure the intensities from a sample, but do not know what the composition is (and for coatings and

67

other heterogeneous samples, we also may not know the critical dimensions). The only practical way to use Monte-Carlo in these situations is to establish calibration curves for semi-routine use (an example is shown in Figure 5.15), or to refine analytical expressions used for traditional calculations and corrections.

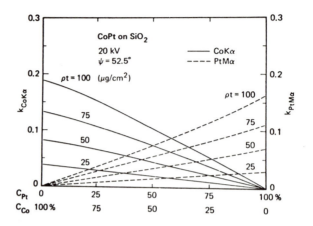

Figure 5.15 Monte-Carlo calculated calibration curves for Co and Pt X-ray emission from thin films on a SiO_2 substrate (Kyser, 1976).

Monte-Carlo simulation remains a powerful research tool for study of the underlying physics of these techniques, and a marvelous teaching aid to help us visualize the processes, but will only rarely be the method of choice in routine analytical situations.

Chapter 6

X-Ray Absorption

We have seen that the X-rays produced in the sample by ionizatioɴ. of atoms struck by the high energy electrons, have a distribution in depth. This means that most of them will have to pass through some of the sample to get out, and during that passage they will suffer absorption. Whenever X-rays pass through anything, they are absorbed according to Beer's Law:

$$I = I_o \, e^{-\mu\rho x}$$

where I is the intensity (number of photons) left, I_o is the original intensity, ρ is the density and x the distance, and μ is the "mass absorption coefficient". This depends on the element(s) in the material and on the energy of the X-rays. It combines several physical processes which together serve to reduce the X-ray intensity exponentially with distance (as shown in Figure 6.1).

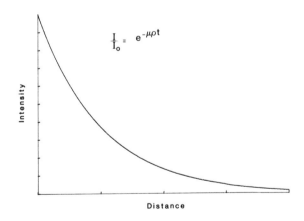

Figure 6.1 Exponential decrease of X-ray intensity with distance.

The most important interaction (from our perspective of X-ray analysis, and in terms of magnitude of the effect) is the absorption of photons to create photoelectrons and leave (briefly) ionized atoms in the sample. The probability of this occurring is greatest when the photon energy is just above the binding energy for a particular shell, and falls off exponentially as the photon energy is increased above that value. Of course, if the photon energy is below the binding energy, it cannot be absorbed at all by that shell (although it may still be absorbed by the next shell out, which has a lower binding

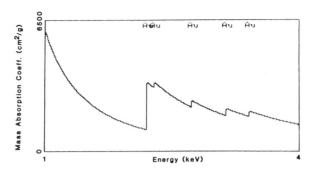

Figure 6.3 Mass absorption coefficient for gold as a function of energy, with discontinuities at each of the five M absorption edges.

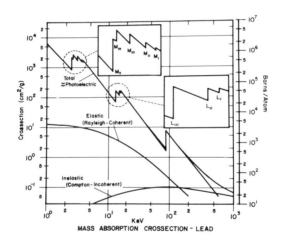

MASS ABSORPTION CROSSECTION - LEAD

Figure 6.4 Mass absorption coefficient for lead, showing the elastic and inelastic contributions and the K, L, and M edges (Woldseth, 1973).

    The general trend of each curve shows a rise as energy decreases, with sharp discontinuities at the energies of the shell binding energies (these are in fact often referred to as the element's absorption edge energies). This is in agreement with the description given above of likelihood of absorption being greatest when the photon has an energy close to (and greater than) the edge. Plotted on a log scale these are nearly straight lines, and show that the contribution of each additional shell is added to the lower energy shells as the binding energy is passed. In other words, the total mass absorption coefficient represents a sum of the absorption probabilities of each shell with a binding energy below the photon energy, and for each shell the probability is closely approximated by a function with a term decreasing exponentially with energy difference above the edge.

71

Collections of mass absorption coefficients have been obtained by measurement, and fitted to convenient equations for calculation. Heinrich's model (1966) is most often used in this field, and is shown below: $\mu(Z,E)$ where Z is the absorbing element and E the photon energy is

$$A\ e^{(B\log^2(Z)+C\log(Z)+D)}\ x\ (12.398/E)^F$$

where A, B, C, D and F are constants (or nearly so) as listed below. You must select the set of numbers for the highest energy edge of the element that is less than the photon energy E.

Shell	A	B	C	D	F
N	1.00	1.359165	-9.492116	18.64081	2.22
M5	0.5609	0.2562163	1.15119	-5.684848	2.60
M4	0.6169	"	"	"	"
M3	0.7143	"	"	"	"
M2	0.8621	"	"	"	"
M1	1.00	"	"	"	"
L3	0.6135	-0.2544711	4.769245	-10.37878	2.73 (*)
L2	0.8547	"	"	"	"
L1	1.00	"	"	"	"
K	1.00	-0.2322294	4.070053	-6.220746	(**)

(*) for Z > 42 use F =

$$\exp(-0.113159\log^2(Z)+0.8368829\log(Z)-0.5459687)$$

(**) use F =

$$\exp(-0.0045522\log^2(Z)-0.0068535\log(Z)+1.070181)$$

This expression, which is actually rather straightforward in spite of the mass of numbers, is readily adapted to small computers and can be used to compute $\mu(E,Z)$ when needed. Of course, most materials of real interest are not single elements, so we must sum up the individual element mass absorption coefficients to obtain the total matrix absorption coefficient. The C values in the expression below are the concentrations expressed as weight fraction:

$$\mu(matrix,E) = \Sigma\ (\mu(Z,E)\ C_Z)$$

This means that in a complex matrix the absorption coefficient for each energy X-ray can be calculated only by knowing the matrix composition, and that it will vary with energy following a curve that has discontinuities at the absorption edges of each element present. Of course, the magnitude of the steps will be small for the minor elements. Figure 6.5 shows a the plot of mass absorption coefficient vs. energy for a stainless steel containing 18%Cr, 9%Ni, 1%Mn. The Mn edge is practically invisible, as we would expect. The vertical scale on the plot is in $cm^2/g.$, the usual units for mass absorption coefficient. Remember that the intensity decrease is exponential with the product $\mu\ (cm^2/g)$ times ρ (density in g/cm^3) times distance (cm).

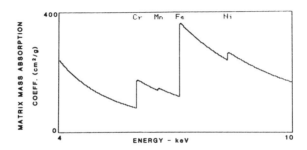

Figure 6.5 Mass absorption coefficient as a function of energy for stainless steel, showing the contribution of each element's absorption edge.

The Depth Distribution and $F(\chi)$

As simple as absorption is to calculate, it becomes enormously complicated for the case of the X-rays generated in the sample by the electron beam. This is because of the depth distribution of the X-rays, which means that the total emitted (as opposed to generated) intensity is an integral:

$$\int_0^{range} \phi(\rho z)\ e^{-\mu\rho z\ \csc(\theta)}\ d(\rho z)$$

where the range of the integral is the total electron range as described previously, and μ (mass absorption coefficient), ρ (density) and ϕ (the depth distribution of generation) have their usual meanings. The angle (θ) is called the X-ray takeoff angle, and is defined as shown in Figure 6.6. Note that for a given depth z, the path length of the X-rays that reach the detector is increased by the cosecant of the angle.

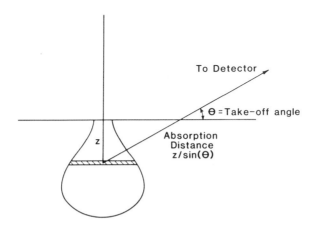

Figure 6.6 Schematic diagram of X-ray generation in sample and the absorption of photons from each depth z enroute to the detector, depending on the X-ray takeoff angle θ.

73

The measured intensity will be this integral multiplied by the weight fraction of the element in the matrix and by the solid angle covered by the detector ($\Omega / 4\pi$) since the characteristic X-rays are generated isotropically (uniformly in all directions). We shall omit the detector angle from the equation since it is presumeably an instrument constant, but include the element's composition (weight fraction) C. It is also useful to substitute the value χ for the quantity μ csc (θ). This gives us:

$$I = C \int_0^{inf.} \phi(\rho z)\, e^{-\chi(\rho z)}\, d(\rho z)$$

which mathematically is the familiar form of the Laplace transform. If the function $F(\chi)$ can be found which is the Laplace transform of the function $\phi(\rho z)$, then the intensity will simply be:

$$I = C\ F(\chi)$$

There has been a considerable amount of searching for functions to use for $\phi(\rho z)$ that satisfy two criteria: 1) that they are integrable and have Laplace transforms, and 2) that they fit, or at least approximate the shape of the actual $\phi(\rho z)$ distribution curve. Note that we are in some danger in looking at $F(\chi)$ functions in losing touch with the physical basis for our understanding. χ is the product of μ and the cosecant of the takeoff angle, a sort of 'absorption magnitude'. $F(\chi)$ will represent the fraction of the generated X-rays that are detected, as a function of this (χ) term.

Several functions have been proposed for $\phi(\rho z)$ that suit the first criterion above (having a corresponding $F(\chi)$ equation), but with varying degrees of success as far as the first criterion (matching reality) is concerned. Consider as examples the three functions shown schematically in Figure 6.7:

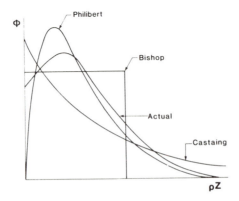

Figure 6.7 $\phi(\rho z)$ curves used to model the depth distribution of X-ray generation.

Superimposed on the shape of an actual ϕ (ρz) distribution are the models proposed by Castaing (1951), Philibert (1963), and Bishop (1974). There have been others; these were selected to illustrate several specific points. Notice that the Castaing curve:

$$\phi = e^{-k\rho z}$$

is a simple exponential distribution that has the greatest generation right at the surface, not at all like the real curve, and also extends to infinity in depth while the real ϕ (ρz) curve has a definite limit. The Philibert model, on the other hand, is:

$$\phi = 4 \; (e^{-\sigma\rho z} - e^{-\sigma\rho z(1+1/h)})$$

where h is Z^2/A (Z=atomic number, A=atomic weight), and σ is the Lenard coefficient that describes the number of electrons in the sample at a depth ρz as declining exponentially as $\exp(-\sigma\rho z)$. The value of σ may be calculated using an expression of the form shown below, in terms of the initial electron voltage and the critical excitation energy of the element being analyzed, with constants as determined by two authors:

$$\sigma = K \; / \; (E_o{}^n - E_c{}^n)$$

1) Duncumb & Shields (1966)

$K = 2.39 \; 10^5$
$n = 1.5$

2) Heinrich (1970)

$K = 4.5 \; 10^5$
$n = 1.67$

The latter is more used at this time. This expression of Philibert (1968) is zero at the specimen surface (there is a more elaborate form that does not have this constraint), again quite different from the physical case. The Laplace transform it produces is:

$$F(\chi) = 1/((1+(\chi/\sigma))(1+(\chi/\sigma)(h/(1+h)))$$

The Bishop model is a simple step function, as though X-ray generation was constant with depth for a certain depth z' and then dropped abruptly to zero. This produces a Laplace transform:

$$F(\chi) = (1-e^{-2\chi z'}) / 2\chi z'$$

$z' = (2h+1)/(\sigma \; (h+1))$ where $h = 1.2 \; A/Z^2$
and σ is as used above.

We have seen that there are indeed some functions that have Laplace transforms but seem to offer exceedingly crude approximations of the depth distribution $\phi(\rho z)$. The equations for $F(X)$ are manageable, but have lost any possible intuitive meaning. Furthermore, they look very different from each other. Which (if any) should we use?

If we actually plot the various $F(X)$ functions from these and other expressions, we find as shown in Figure 6.8 that they are all very close indeed. At least over the range of moderate X (up to say 5000), the F value (greater than 0.7) can be obtained using any of the expressions. This means that when absorption is not too great (X is large when mass absorption coefficient times distance is large, so either very absorbing matrices or low takeoff angles can produce large X), there is no sensitivity in the $F(X)$ function to the actual form of $\phi(\rho z)$.

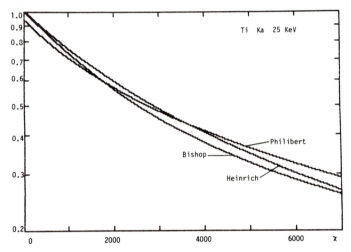

Figure 6.8 $F(X)$ curves comparing different models.

Even though we had approximations that should clearly have underestimated and overestimated the absorption, the resulting $F(X)$ curves agree rather well so long as the absorption parameter "X" is not too large. And when they do diverge, it is as expected, with the Bishop model showing a greater correction (more absorption) then the Philibert, so we can still be confident that the 'right' answer, the unknown $F(X)$ curve that corresponds to the true $\phi(\rho z)$ distribution, will lie within the spread of the various $F(X)$ curves.

The next step in the evolution of $F(X)$ functions was to fit simple equations to the various $F(X)$ functions without regard to the existence of a Laplace transform (ie. for which no explicit $\phi(\rho z)$ curve exists). One approach was used by Anderson and Wittry (1968), who were able to transform the $\log(F(X))$ curve to a linear function of X' where X'/X is:

$$AZ^{-4/3} (E_0^{1.8}/(1.8\log(174E_0/Z))-E_c^{1.8}/(1.8\log(174E_c/Z)))$$

76

Another very satisfactory expression, used in many correction programs, is Heinrich's (Yakowitz, 1973):

$$1/F(X) = (1 + 1.2\,10^{-6}\ (E_o^{1.65} - E_c^{1.65})\ X)^2$$

in which terms very similar to our previous expressions for electron range appear (not surprisingly, since they describe the depth of penetration, which in turn controls the amount of absorption). For values of X less than about 4-6000, this fits all of the $F(X)$ curves well enough to be used with confidence to estimate quantitatively the amount of absorption suffered by the X-rays generated in depth in a sample. This can also be expressed in terms of the value of F itself: if $F(X)$ is above 0.8 (20% absorption) the value is OK, but below 0.6 (40% absorption) the simple expression may not be adequate (it is in this range that the multitude of $F(X)$ curves diverge).

Inclined Surfaces

If the specimen surface is not perpendicular to the electron beam, the situation is altered. The roughly spherical excited volume is no longer symmetrically aligned under the impact point of the electron beem (at least as viewed from the surface), and the perpendicular depth below the surface for X-ray generation is decreased. We have already seen, in the discussion on electron penetration, that the shape of the $\phi(\rho z)$ curve, the amount and energy distribution of the backscattered electrons, and the total amount of X-ray generation is also affected by non-normal electron beam incidence. Depending on the location of the X-ray detector with respect to the direction of tilt, the absorption path length that X-rays must traverse may be increased or decreased.

When the tilt is directly toward the detector, we can draw the geometry rather simply in the plane defined by the detector and the electron beam. Figure 6.9 shows that for a tilt angle B and a detector takeoff angle θ relative to the horizontal plane through the

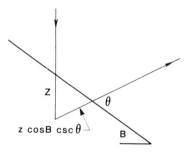

Figure 6.9 Simple geometric correction for absorption path length when specimen tilt is toward detector.

point of intersection of the beam with the specimen, the absorption path length is changed in proportion to the cosine of B. This can be taken into account straightforwardly by using for χ the modified value:

$$\chi = \mu \cos(B) \csc(\theta)$$

This simplistic approach is more complicated if the direction of tilt is not directly toward the detector. It is then necessary to specify an additional angle, called the Azimuth angle "A", which is the angle in the horizontal plane between the direction of tilt and the plane containing the electron beam and the detector. Figure 6.10 shows the three-dimensional geometry. Applying trigonometry to this configuration allows calculation of the absorption path length in terms of the angles, and the modified value of χ becomes (Moll, 1977):

$$\chi = \mu \cot(B) \, / \, (\cos(A) \cos(\theta) + \sin(\theta) \cot(B))$$

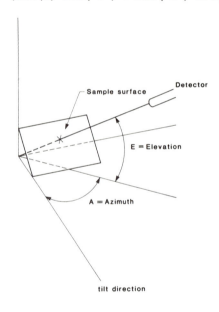

Figure 6.10 Angles A=Azimuth and E=Elevation define the detector location with respect to an arbitrarily tilted surface.

Many modern correction programs incorporate this term to modify the absorption correction $F(\chi)$. This certainly gives improvement over ignoring the tilt and considering the surface to be flat, since it can change χ by a factor of 2 or more. However, it is somewhat too simplistic an approximation. The lateral distribution of X-ray generation (which is modified slightly as compared to that for normal incidence) places some of the X-rays generated at a given depth in the specimen below the impact point quite close to the tilted surface and others much farther away. The change in absorption path length that we have calculated applies only to points along the midline of the distribution. Since absorption is an exponential, and not a linear function of distance, the average absorption path length from the center of the distribution does not give the average absorption. There will in general be less absorption of X-rays when the tilt acts to reduce the path lengths than we would compute using the average geometric path. In other words, the modified X value as shown above leads to an $F(X)$ that overestimates the actual absorption taking place.

Nor is it possible to simply correct this for the shape of the lateral distribution. In the first place, the lateral spread is a function of atomic number (which controls scatter angles) and so we cannot expect to find a universal function. The most common expressions for $F(X)$ do not include a specific dependence on Z. Additionally, the lateral distribution has not been modelled analytically, but is only obtained by a Monte-Carlo calculation.

Indeed, for the case of a tilted specimen surface, it is only by carrying out a Monte-Carlo calculation for electron trajectories and then applying a discrete absorption calculation for each generated X-ray along a path leading to the specified detector position that an accurate compensation for absorption can be obtained. As we have already seen, this is not a suitable tool for routine analysis of unknown samples. However, it is an excellent way to verify more approximate models that can be applied, by comparing the results for known specimen compositions and analysis conditions. Provided the total $F(X)$ correction is not too large (greater than 0.7), which reflects the size of the mass absoption coefficient and the nominal absorption path length, and the tilt angle is not too great (less than about 40 degrees), the simple model for altering X as a function of the angles A and B is acceptable. When the tilt angle is above about 60 degrees or the average matrix atomic number is high (which significantly changes the shape of the excitation volume), the approximation is poor. Under these conditions, we have already seen that the correction for excitation loss due to backscattering is also questionable.

As a final point, it is necessary to emphasize out that the angles A and B, which define the orientation of the sample surface with respect to the detector position, are not always easy to obtain. For a perfectly flat specimen mounted with precision on the sample stage, the angles could be read from, ar at least calculated from the settings of the stage control knobs. This situation is rarely encountered in the SEM, since by its nature it is usually employed to deal with samples having irregular surfaces. In the electron microprobe, where flat, polished surfaces are used, the surface is likely to be normal to the beam, or at least in a constant position which will also be used for the standards (in which case the magnitude of the errors is greatly decreased, and cancels out if the unknown and standard are very similar).

For a sample with an irregular surface, such as a fracture face, the orientation of a local region can in principle be determined by stereoscopy: combining two images taken at slightly different angles to determine elevations by parallax. Consider for example the situation shown in the drawing of Figure 6.11: the original surface, defined by two recognizeable points in the image, is inclined at an unknown angle B; by tilting it a further amount I, which can be measured, the distance between the two image points decreases. Since their actual distance along the surface remains unchanged, it is possible to solve trigonometrically for the original unknown angle B in terms of the incemental angle I and the measured separation of the points in the two images (this parallax measurement can often be obtained directly from the image screen without even taking photographs).

$$B = \tan^{-1} ((\cos(I) - D_2/D_1) / \sin(I))$$

Figure 6.11 Geometry used to calculate filt angle B from parallax measurements D before and after applying an incremental tilt I.

Unfortunately it requires carrying this procedure out in two directions to completely define the surface orientation. Most SEM stages do not have the capability to tilt the sample in two orthogonal directions, and so the only recourse is to shift the specimen in X and Y, which also gives the desired parallax in the image (but which only works at low magnification).

Because of the difficulty of determining surface orientation on irregular surfaces, another technique has come into limited use which performs quantitative calculations including the absorption correction without needing the local tilt or takeoff angle. It will be discussed more fully later on, but in brief consists of measuring two sets of elemental intensities from the same position on the sample, but with different accelerating voltages. Since this changes the depth of penetration, it also changes the absorption of the X-rays (and of course the equations we have seen for $F(\chi)$ include E_o the accelerating voltage). The change in absorption (as well as other parameters) affects the different elements present to varying degrees. There will only be a single surface orientation for which both sets of data give the same calculated sample concentration, and this is found by an iterative scheme in which an arbitrary takeoff angle is assumed, calculation of concentrations is carried out, and the assumed angle is varied to minimize the difference between the compositions computed from the two sets of data. The final result is both the sample concentration and, as a by-product, the local surface orientation. Since conventional expressions for $F(\chi)$ and other parameters are used, the usual constraints on accuracy apply, with the exception that at large tilt angles the error due to lateral distribution will cause the estimated angle to be wrong, but as the error appears in both sets of calculations in the same way, the final error in the concentration is not so great (the individual errors tend to cancel out).

Chapter 7

Secondary Fluorescence

The process of absorption just described reduces the intensity of the X-rays from one element, with the principal mechanism being the photoelectic ionization of atoms in the matrix. Many of these ionizations will involve inner shells, provided the energy of the X-ray being absorbed is greater than the various elements' binding energy. Some of the excited atoms will then emit characteristic X-rays of their own, which will add to the intensity from that element produced by direct electron beam excitation. This process is called secondary fluorescence.

In principle, all of the radiation in the sample whose energy is above the absorption edge energy of the element of interest is capable of fluorescing the element, and creating additional emitted X-rays. As shown in Figure 7.1, this would include the characteristic lines of other elements, both the major lines (α-1) usually measured for analysis and the minor lines often ignored but still present in the spectrum, as well as the continuum or Bremsstrahlung. An integration of all of this intensity (corrected to reflect its own absorption, so that it becomes a spectrum of generated intensity) multiplied by its ability to ionize the analyte element, can in principle be performed. In practice, it is simpler to add the separate contributions of the characteristic lines of the major elements present, often ignoring altogether the minor lines and the continuum. In the discussion which follows, a number of approximations are made in order to simplify the calculation (which nevertheless is mathematically the most complex we shall encounter).

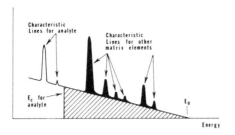

Figure 7.1 Secondary fluorescence of each analyte is produced by all X-rays generated in the sample with energies higher than the absorption edge energy of the analyte.

Our first assumption is that one half of all the emitted primary radiation leaves the surface of the sample, and one half penetrates into it and is absorbed. This is justified on the grounds that the depth of excitation produced by the electrons is extremely shallow, and that the range of the X-rays is much larger, so that on the dimension of the X-ray penetration, the original region of excitation is effectively a point at the specimen surface. The factor of 0.5 is adequate for this purpose, but it is interesting to note that this means that the volume of the sample which can contribute secondarily fluoresced X-rays is much larger than the original excited volume, and compositional inhomogeneities over a much greater scale can affect the results. Also, if the sample is not a flat-surfaced semi-infinite solid, but has some other shape, say a spherical or cylindrical particle or a thin section, a much higher fraction of the primary radiation will escape and reduce dramatically the magnitude of the secondary fluorescence effect.

Long range secondary fluorescence effects can reach across phase boundaries, to alter the apparent compositions and even generate the appearance of significant concentrations of elements not present in the region of primary excitation at all. Composition profiles measured along a line across a phase boundary can, with difficulty, be corrected for these effects. The analysis of small regions, of the order of the size of the excited volume, is possible only if secondary fluorescence of the elements in the surrounding matrix is small or absent.

In a sample containing the analyte element i with concentration (weight fraction) C_i, the fraction of the X-rays absorbed by element i is:

$$C_i \mu_i(E_j) \; / \mu_{matrix}(E_j)$$

where the mass absorption coefficient values are evaluated for the energy E_j of the fluorescing photons (usually this will be the energy of the emission line from a second "fluorescing" element designated here as "j"), and the matrix mass absorption coefficient is a sum of the coefficients for each element in the matrix times the element's weight fraction. Note that as in many of these corrections, we need to know the entire sample composition in order to compute the inter-element effects.

Of the total absorption in element i, the fraction that is absorbed by a particular shell (for instance the K shell) is given by the relative height of the step in the mass absorption coefficient curve at the edge for that shell. This quantity is called the absorption edge jump ratio, and is defined as shown in Figure 7.2; "r" is the ratio of the mass absorption coefficient on the high energy side divided by that on the low energy side.

The value we actually want is $(r-1)/r$, the fraction of the absorption that is accounted for by the shell. For a given shell, this value varies in a gradual and smooth way as a function of Z (except when the next shell out is partially unfilled, as for the K lines of ultra-light elements). Measured data have been fit to simple expressions which are more easily computed than determining the mass absorption coefficients on both sides of the edge for every fluoresced

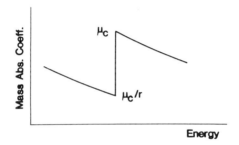

Figure 7.2 An idealized absorption edge showing the jump ratio r.

element. For the K and L III (which corresponds to the L- α -1 line used most commonly for analysis) shells, this quantity is given approximately by:

$(r-1)/r$ for K shell$=1.15265$ $z^{-.083868}$

for L shell$=0.95482-2.5925$ 10^{-3} z

Some calculation programs simplify this even further and use constant values, 0.88 for K and 0.75 for L.

The number of additional ionizations of the fluoresced element which produce characteristic X-rays is the fluorescence yield, which was previously described. Now we require equations with which to evaluate it. Several have been proposed, as listed below. The Burhop expression is probably the most widely used, and is more accurate than the simpler Wentzel equation because it includes screening effects of one shell on another and relativistic effects. In both cases the functional form is derived from theoretical considerations and the constants from measurements. The other equations are chiefly just convenient forms to fit to measured data, and often do not apply to all of the K,L,M shells of interest.

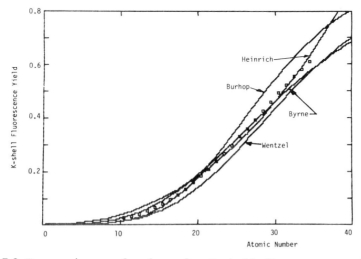

Figure 7.3 Expressions and values for K-shell fluorescence yield.

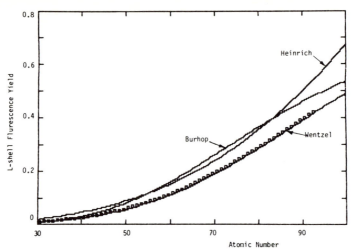

Figure 7.4 Expressions and values for L-shell fluorescence yield.

1)Wentzel (1927):

$$\omega = Z^4 / (a + Z^4)$$

$$a(K) = 1.15 \ 10^6$$

$$a(L) = 1.0 \ 10^8$$

$$a(M) = 7.5 \ 10^8$$

2)Burhop (1955):

$$(\omega/(1-\omega))^{1/4} = A + BZ + CZ^3$$

	A	B	C
K-shell	-0.03795	+0.03426	+1.163 10^{-6}
L-shell	-0.11107	+0.01368	-2.177 10^{-7}
M-shell	-0.00036	+0.00386	+2.010 10^{-7}

3)Byrne&Howarth (1970):

$$\log((1-\omega_K)/\omega_K) = 11.66 - 3.36 \log(Z)$$

4)Heinrich (Yakowitz, 1973):

$$\omega = \exp(A \log(Z) + B)$$

	A	B
K-shell	2.373	-8.902
L-shell	2.946	-13.94

84

Figures 7.3 and 7.4 compare these expressions for the K and L shell fluorescence yields to the data published by Burhop and Assad. The agreement is imperfect, but adequate for use in fluorescence corrections, particularly considering the other sources of error and the generally small magnitude of the total correction.

It is worth noting here that the L-shell fluorescence yield is actually that for the LIII shell, from the excitation of which is emitted the L- α -l line normally used for analysis. The ratio of yields for the other shells to this one is not constant, but varies somewhat from element to element. These internal or Coster-Kronig transitions add complexity to analysis using L lines where different line intensities are needed (and also in cases where L peaks overlap and must be stripped). They are not well measured, modelled, or included in many correction procedures. The situation would be even worse for the M lines, except that they are less used for analysis, and other errors are even worse in their final effect.

If we combine the expressions so far described for the additional flourescence intensity from element "i" due to radiation from element "j" we have:

$$I_j = 0.5 C_i (\mu_i (E_j) / \mu_{mtx} (E_j)) ((r_i - 1) / r_i) \omega_i I_j$$

showing a dependence upon the parameters μ, r, and ω which we can understand and compute, a proportionality to the concentration of element i (we must have enough atoms to absorb the fluorescing radiation), and the intensity of the fluorescing element. The latter quantity is inconvenient, and we would like to replace it with some function of the elemental concentration, so that the fluorescence contribution can be more directly evaluated. This requires that we have an expression for the relative intensities of elements j and i, in terms of their concentrations.

For two elements in a sample, the relative generated intensities will be proportional first of all to the relative concentrations. But we have so far found it convenient to express concentration in terms of weight fraction (for example to use in summing up absorption coefficients), and in this context we are dealing with the relative atomic abundance of the two elements. Hence we shall have to use C/A where C is the weight fraction and A the atomic weight. These are mostly available in tabular form in present day programs, but in an earlier period when even that amount of computer memory (100 real numbers will consume about 400 bytes of storage) was expensive and jealously guarded, the equation shown below was used to approximate A as a function of Z (it may still be useful to keep it with your pocket calculator). Figure 7.5 shows the overall agreement of the function with true values, and the notable outliers for a few elements.

$$A = 1.428 Z^{1.1289}$$

In addition to the relative atomic abundances of the two elements, the relative excitation, which depends on the excitation voltage E_o must be considered. It has proved generally acceptable to use as an approximate model for relative cross section:

$$(U-1)^{5/3}$$

where U is the overvoltage introduced previously. We will discuss the variation of cross section with E and Z in more detail later on, when considering standards, but this expression is adequate at least for the fluorescence correction. The ratio

$$((U_j-1)/(U_i-1))^{5/3}$$

gives the relative excitation of two elements provided the same shell is involved for both. The constant terms, which have to do with the electron density in the shells, cancel if both the fluorescing line and the fluoresced line are both K, or both L. If they are not, then an additional constant must be introduced. The best empirical value for this factor is 0.24 for K fluorescing L and its reciprocal, 4.2, for L fluorescing K. Fluorescence by M lines is generally of minor importance, and is ignored in many programs; similarly the use of M lines for analysis (and consequently the need to perform a fluorescence correction on them) is discouraged for reasons of accuracy of many other terms and parameters, as will be discussed later.

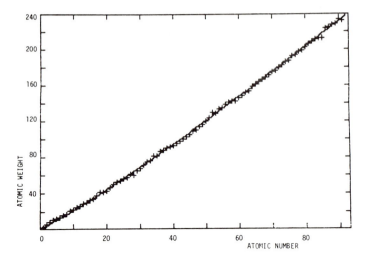

Figure 7.5 Variation of atomic weight with atomic number.

In addition to the ratio of relative excitation of the fluorescer and analyte elements, the fluorescence yield must be included for each to give the number of X-rays generated by the ionizations. If all these terms are included, the equation for additional generated intensity from element "i" becomes (Reed, 1965):

$$0.5C_j\,\omega_j(\mu_i(j)/\mu_{mtx}(j))((r_i-1)/r_i)(A_i/A_j)P_{ji}((U_j-1)/(U_i-1))^{5/3}$$

In this expression, we find the concentration of the fluorescing element, its fluorescence yield, the mass absorption coefficients for the fluorescing radiation in the analyte element and in the matrix as a whole (which depends on the concentraton of all the elements present), the absorption edge jump ratio for the analyte, atomic weights, the P factor which is 1.0 for K-K and L-L fluorescence but is taken as .24 for K-L and 4.2 for L-K, respectively, and the relative cross sections of the elements calculated from their respective overvoltage ratios. Other formulations have been used for this term, particularly involving different models for cross section and relative excitation, but without showing any significant improvement.

The fluorescence radiation is subject to absorption in the matrix, of course (in fact, as it is generated much deeper in the sample that the direct excitation due to the electron beam, the amount of absorption is much greater). The evaluation of the absorption depends on two exponential approximations: the first is that if the primary radiation is produced very near the surface, then the distribution in depth of its absorption (and consequently the distribution of generation of fluoresced X-rays) will be exponential; the second is that the generated X-rays are absorbed exponentially with distance, again with the path length defined in terms of the depth and the cosecant of the takeoff angle (or the equivalent term for non-normal incidence). The first assumption may be questioned if high electron beam voltages are used to analyze elements with low energy X-rays (not a recommended procedure since it produces very large values of X and large total absorpion, and we have seen that $F(X)$ is not accurately computed in this case). In that case, the depth distribution of primary radiation would be great, and the absorption distance of the low energy primary X-rays would be small. However, for most real cases it is an acceptable approximation. The second approximation is fine if the sample is the classic flat surfaced, semi infinite solid, but not for more irregular surfaces.

If the exponentials are used, then a double integration of excitation in depth and absorption of the emission must be carried out. In terms of two parameters y and v, the result is a term to be multiplied by the generated intensity:

$$\log(1 + y) / y + \log(1 + v) / v$$

where $\quad y=(\mu_{mtx}(E_i)/\mu_{mtx}(E_j))$

describes the absorption of the fluoresced radiation and

$$v = \sigma / \mu_{mtx}(E_j)$$

describes the absorption of the primary radiation in depth using σ, the same Lenard coefficient we encountered in discussing the absorption correction for primary radiation in the previous chapter. Modifications of these terms which take into account a more realistic model of the depth distribution of the original radiation have been published and evaluated, but offer negligible improvement at considerable calculational expense.

The fluorescence correction is completely ignorable in many cases. One is clearly the case when the matrix element(s) do not have emission lines above the aborption edge of the analyte. Another is when the other elements' lines are above the absorption edge, but very far away. The magnitude of the correction depends on the ratio of mass absorption coefficient of the fluorescing radiation by the analyte, to the total matrix absorption coefficient. Remembering the exponential drop off of mass absorption coefficient with energy above the absorption edge, we can predict that the X-rays from another element will produce very little fluorescence if they are far from the analyte's excitation energy. Indeed, many programs skip the calculation entirely for elements whose emission energies are more than about 5 keV above the analyte's absorption edge. Similarly, trace elements will not cause significant fluorescence since the contribution is in proportion to their concentration.

The practical calculation of fluorescence corrections is made simply by summing the term shown above (including the absorption correction) for the effect of each matrix element (if it causes significant fluorescence at all). The result depends on the complete matrix composition, and is a number which gives the fractional increase in intensity of the analyte's measured emission line. Hence, dividing the measured intensity by 1.0 plus the sum of the individual terms will remove the effects of fluorescence. This approach is used by most programs and is usually adequate, but it ignores two other sources of fluorescence intensity which can be important in some cases.

First, we have ignored the contribution of emission lines from the matrix elements other than the principal one. This is not too much of a problem when one element's K line fluoresces another's K edge, since the K-α intensity accounts for 85-90% of the total K emission intensity, and the K-β line is not only much smaller than the K-α but is farther away from the absorption edge so that its effect is less. This is clearly not an acceptable simplification in some cases, however. Consider brass alloys (principally copper and zinc). The important line and edge energies for these elements are listed below:

Element	K-edge	K-α	K-β
Cu	8.98	8.04	8.91
Zn	9.66	8.63	9.57

There is no characteristic fluorescence of the zinc, since its critical excitation energy is higher than anything else. Most programs will also decide that there is no fluorescence of the copper, since the zinc K-α line is not above the copper absorption edge. But the zinc K-β line can fluoresce the copper, and indeed do it very efficiently because it is so close to the edge. This can produce errors of a few percent in alloys with high zinc content.

The situation can become much worse when L lines are involved. First, the principal L-α-1 line contains only about half of the total shell emission, so that there may be quite a few other lines able to contribute significantly. Also, the L lines are spread out over a greater energy range than the K lines are, so it is much more common to have a 'minor' line at higher energy and able to produce fluorescence.

88

Another important problem in dealing with L shell fluorescence is the use of a single absorption edge energy for the element (when in fact the relative excitation of each subshell is different, and all contribute to the emission), and a single mass absorption coefficient (again, all the lines have different energies and consequently different absorption coefficients, and in complex matrices there may be edges of other elements occuring between them). Likewise, the absorption edge jump ratio is properly broken down into the three subshell edges, and the fluorescence yield should not be the total shell yield, but should be considered separately for each subshell. In practice, this is rarely done; weighted averages are used, and it is hoped that errors will be small and compensating (especially if the standards used are similar in composition to the unknowns).

Characteristic lines from the major matrix elements present are not the only possible source of fluorescence of the analyte, of course. Continuum or Bremsstrahlung radiation is produced in the sample, and those photons which are just above the absorption edge should be able to produce additional ionizations. If a distribution of continuum intensity following Kramers' law is assumed, equal to $aZ(E_o-E)/E$ as a function of energy E and the electron accelerating voltage E_o, then it is possible to integrate the contribution of all energies from the absorption edge energy E_c up. The integration, combined with the same kinds of assumptions used in dealing with characteristic fluorescence, gives the following expression for the generated fluorescence radiation intensity due to the continuum:

$$0.5a\omega(r-1)/rZE_c(U\log(U)-U+1)$$

where ω (fluorescence yield), r (absorption edge jump ratio), Z (atomic number) and U (overvoltage ratio E_o/E_c) have their usual meaning. To judge the magnitude of this term, let us use 0.8 for $(r-1)/r$, introduce the approximate value of 2×10^{-9} for Kramers' constant "a", incorporate the simplest expression for ω, and use Moseley's law in which E_c is roughly proportional to Z^2. We also need a rough estimate of the primary intensity in absolute units, which we shall borrow for now from a future chapter. Combining everything we find that the ratio of fluorescence intensity generated by the continuum to characteristic intensity is roughly estimated by:

$$10^{-7}Z^4$$

which means, quite simply, that for low average atomic number matrices the effect is easily and justifiably ignored, but for high Z samples it may become significant. The expression predicts only a 2% increase in intensity for Z=21, rising to 5% for Z=27, 10% for Z=32, 20% for Z=38 and so on. In particular, the analysis of light elements (low excitation edge energy, low primary characteristic intensity) in a heavy matrix can show a significant effect. Even more significant is the error introduced by using pure heavy element standards (in which a significant fraction of the measured intensity is due to continuum fluorescence) for low concentrations of the same elements in a low Z matrix, where the total continuum produced is much less. Also, the presence of relatively high Z particles in a low Z matrix (often found in stained biological materials) can produce significant amounts of

Bremsstrahlung from scattered electrons which strike them, producing fluorescence of large regions of the sample, but this goes beyond the meaning of a normal 'fluorescence correction' and becomes more of a problem of stray signals. Usually it is hoped that continuum fluorescence will cancel out between standard and unknown if they have about the same composition and hence the same average atomic number.

Our discussion of secondary fluorescence has so far also ignored the possibility that the absorption of fluoresced radiation may give rise to still further ionizations and produce tertiary fluorescence. A classic example of a case in which this can occur is the analysis of stainless steels. The principal elements are listed below with their important energies:

Element	K—edge	K—α
Cr	5.99	5.41
Fe	7.22	6.40
Ni	8.33	7.47

The nickel X-rays can fluoresce the iron and the chromium, the iron can fluoresce the chromium, and hence it is possible that iron radiation fluoresced by the nickel may produce additional chromium radiation. Detailed calculation of these terms shows us (fortunately for the simplicity of our every-day programs) that this is not an important concern even in this, nearly worst case. Depending on the particular composition chosen, the fluorescence term for Cr will be about 20%, of which 18 will be due to the iron, 2% to the nickel, and less than .05% due to the tertiary process of nickel to iron to chromium.

The oversimplification of the fluorescence correction is justified on the grounds that it makes the calculations practical for use in smaller computers (a weak excuse as they become ever less expensive), and that it gives acceptable accuracy in most cases. The latter is particularly so if the standard used is similar to the unknown, so that the errors will be similar in sign and magnitude and will tend to cancel. Also, the other sources of errors in L or M line analysis, particularly the uncertainty in mass absorption coefficients and fluorescence yields, are often larger than the potential error in the oversimplification in fluorescence correction. As more measurements are made to refine fundamental constants, and as microanalysis is performed on increasingly exotic materials where ideal standards are unavailable, and which often involve one or more high atomic number elements (so that continuum fluorescence becomes significant and L or even M lines must be used for analysis), it will become necessary to employ more of the correction terms, expressed more rigorously (Heinrich, 1972).

Chapter 8

Applying the Corrections

(Quantitative Analysis)

The underlying theme of the past few chapters has been not simply understanding the processes that occur in the ionization of atoms by electrons, the emission of X-rays, and their subsequent passage through the sample, but to express that understanding in the form of equations. As Lord Kelvin said:

> "...when you can measure what you are speaking
> about and express it in numbers, you know
> something about it; but when you cannot express
> it in numbers, your knowledge is of a meagre
> and unsatisfactory kind; it may be the beginning
> of knowledge, but you have scarcely
> in your thoughts advanced to the state of
> science, whatever the matter may be..."

Accordingly, we have proposed equations, some theoretically based and some empirical, some quite accurate and others intentionally approximate, to relate the intensities we can measure of the X-rays from elements in the sample back to the concentrations of the elements. Now it remains to put those together to perform the computation, and get that composition.

The most common approach for quantitative electron-excited analysis is known as a "Z-A-F" calculation. The name comes from the "atomic number" (Z), "absorption" (A) and "fluorescence" (F) corrections which are calculated separately pretty much in the same fashion as we have reviewed them here. The "Z" correction is combination of the stopping power and ionization cross section, and the backscattering correction, both of which are primarily dependent on atomic number. Of course, we know that it is the stopping power which determines the depth distribution of generation, that in turn determines the magnitude of the absorption correction, which in its turn gives rise to fluorescence. It is a useful mathematical device, however, to apply these terms individually.

The first step in estimating the concentration of an element in a specimen is simply as the ratio of the measured intensity from the sample to the intensity that would be measured on the pure element. In the next chapter, more attention will be given to the question of getting the standard intensity. For now, we may assume that it is indeed measured from the pure element under identical conditions of accelerating voltage, beam current, surface orientation and so on. It is logical to expect that if a sample gives a measured intensity of, say, 5000 counts per second for iron, and pure iron gives 7000 counts per second, then the approximate iron concentration (in weight percent) should be 5000/7000 or about 71%. In many cases, this first approximation is good enough to serve as a "semiquantitative" analysis without further correction, and in any case it represents the starting point for the Z-A-F process. The ratio of intensity to pure intensity is called the k-ratio, or just k, for the element.

The Z-A-F model adjusts the k-ratio for the effects of other elements in the sample to get the concentration:

$$C = k / (Z A F)$$

The individual terms are as follows: The Z term is actually the area under the $\phi(\rho z)$ curve, which depends largely on atomic number Z. Because this can be discretely integrated, once it is computed, including the absorption effect for each different depth, we can also see that methods using $\phi(\rho Z)$ curves are suitable tools for quantitative analysis corrections. Indeed, they are preferred for very light elements where absorption is the paramount correction. Usually, however, the Z term is calculated as the ratio of R/S for the unknown to R/S for the pure standard. R is just the backscattering correction or effective current factor discussed in an earlier chapter. It is a number less than one which gives the fraction of the incident electrons' energy that remains in the sample and can produce ionization, and can be calculated as a function of Z (of the matrix as a whole) and U (the overvoltage ratio for the element). S is called the "stopping power correction" and gives the total excitation of the element resulting from the ionization cross section and stopping power of the element as a fraction of the total matrix. We saw before that the stopping power was a function of electron energy E and atomic number Z. The total number of ionizatons produced, per electron, in the sample can be obtained by integrating:

$$N_{Av} \rho C_i / A_i \ _{E_o}\int^0 Q \ dE \ / \ (-dE/dx)$$

where the terms outside the integral are Avagodro's number, the sample density and elemental concentration (since C is weight fraction, the atomic weight A is included to obtain atomic concentration) and give the number of atoms of the element present. The stopping power $(-\rho \ dE/dx)$ is taken as before from the Bethe law:

$$2 \pi e^4 N_{Av}(Z/A)1/E \log(1.166E/J)$$

and the ionization cross section Q is computed as a function of overvoltage U and energy E as:

$$0.76 \ \pi \ e^4 \ Z \ \log(U)/ \ (E \ E_o)$$

Note that the fundamental constants cancel out in the ratio of these terms, leaving just the dependence on voltage and atomic number. By introducing the convenience variables M and V:

$$M = \Sigma \ (C_i \ Z_i \ / \ A_i)$$

$$V = \exp \ (1/M \ \Sigma \ (C_i \ Z_i/A_i \ \log(1.166 \ E/J) \ ))$$

where the sums are over all elements in the sample, the integral can be reduced to give for the total number of ionizations (Reed, 1975a):

$$1/2 \ C_i/A \ Z \ b/M \ (U-1-\log(V)/V) \ (EI \ (\log(UV)) - EI(\log(V))$$

where b is a constant and EI is the exponential integral:

$$EI \ (X) = 0.57722 + \log(X) + \sum_{n=1}^{\infty} \ . \ X^n/ \ (n \ n \ )$$

The evaluation of this expression requires a computer, of course. A simpler expression for the total number of ionizations as a function of element and voltage has been introduced by making the approximation that the stopping power can be taken as a constant, evaluated at the average energy (between the initial electron voltage E_0 and the excitation energy for the element E_c. The energy-dependent term then comes outside the integral, which will cancel from ratios between standard and unknown. This makes it possible to simplify the total correction for relative ionization to:

$$2Z_j/A_j/(E_0+E_c \ \log(583 \ (E_0+E_c)/(9.76Z_j+58.5Z_j^{-0.19})))$$

Note that in this expression, the excitation energy is for the particular analyte element, but the calculation must be carried out for each matrix element j and the total "stopping power term" obtained by summing the individual values times each elements concentration (weight fraction). The particular numerical constants in this expression come from the use of the Berger and Seltzer expression for the mean ionization potential J. The approximation made for the stopping power is not unreasonable provided that the accelerating voltage is reasonably close to the excitation energy. If it is too large, the correction will in general be too large because the stopping power is overestimated by the value at the average energy. However, the greatest error in the case of very high excitation voltages is generally in the absorption term, as we have mentioned before.

With both the S and R terms calculable, the total Z term becomes R/S for the unknown matrix divided by R/S for the pure element. The A term is simply the ratio of absorption in the unknown matrix to that in the pure element, and from our earlier discussion it is clear that this absorption is just $F(X)$ where X is the product of mass absorption coefficient for the particular element's radiation and the absorption path length, which reduces to cosecant of the takeoff angle for perpendicular electron incidence, but is somewhat more complex for the case of general tilted surfaces. Hence the A term is just $F(X)$ for the pure element divided by $F(X)$ for the element in its matrix. Note that the mass absorption coefficient for the element's radiation is obtained by summing in proportion to the concentration (weight

93

fraction) the individual mass absorption coefficents for each element in the matrix.

Finally, the fluorescence correction is simply 1 plus the sum of the fractional additional intensities produced by each individual element in the matrix, as described in the preceding chapter. Since there is no fluorescence in the pure standard (we are neglecting continuum fluorescence), no ratio to the pure standard is needed. Hence, if we can calculate Z, A and F, the measured K ratio can be converted to concentration.

But, this is not so straightforward. Every term we have described depends in one way or another on the composition of the matrix, either to determine average atomic number, or to sum terms such as total stopping power, secondary fluorescence or mass absorption coefficients over all elements in the matrix. So even to compute the concentration of a single element in the sample requires that we know the concentration of all the elements. This kind of circular process requires an iterative calculation, carried out by making a first approximation to concentration, computing Z, A and F, applying them to all elements at once to get better concentration values, recomputing the Z, A, and F terms using these second approximations, and so on until "convergence" is reached. This is defined as the point where the same concentrations appear on both sides of the equation, and no further change in the values is taking place (or more realistically, where the difference in values between the n'th and (n+1)'st iteration is less than some preset amount considered to be smaller than the attainable accuracy).

In order to speed up the convergence process, it is common to take advantage of the fact that the shape of calibration curves of intensity versus concentration are at least approximately hyperbolic in shape, that is that (Criss, 1966):

$$(1-k)/k = a(1-C)/C$$

This can be justified on semi-theoretical grounds, and is especially clear for the case of binary alloys (for more complex materials it is useful to think of a psuedo-binary composed of the element of interest and the rest of the sample with all other elements in a fixed ratio). Whatever the net effect of the matrix "element" is on the radiation of the analyte, the effect should be proportional to the concentration (in weight fraction) of the element. Consequently, if we plot the ratio k/C against C, Figure 8.1, we expect to get a straight line (which indeed agrees very closely in all cases and exactly in many situations with what we observe). The departure of the curve from a horizontal line describes whether the net effect of the matrix element is to increase or decrease the intensity of the analyte. Secondary fluorescence can only increase that intensity, whereas changes in average atomic number which alter stopping power or backscattering, and changes in matrix absorption coefficient which may make the matrix more or less absorbing for the element's radiation, can change the slope in either direction. From this linear relationship, the hyperbolic one follows directly.

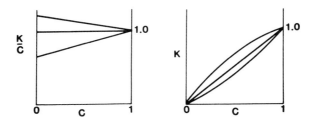

Figure 8.1 The general shape of binary or psuedo-binary calibration curves.

The shape of this curve is defined by three points: the two end points which by definition are fixed at 0,0 and 1,1 and the point C,k corresponding to the current estimate of concentration and the intensity ratio that would be calculated for it (not the observed ratio, for which we do not yet know the corresponding C). Knowing the shape of the hyperbola, the next estimate of concentration C' can be calculated as:

$$k_m C (1-k) / (k_m (C-k) + k (1-C))$$

where k_m = C ZAF based on the previous estimate for C and the corresponding ZAF factors. Practically all present day Z-A-F programs use this convergence process, which has proven to be extremely fast (reducing the number of iterations to between 2 and 4 in practically all cases) and reliable. Many programs also re-normalize the concentrations for all the elements in the matrix to total 100% at each intermediate step in the process, although this can sometimes mask problems caused by missing elements or gross errors in data, which would otherwise manifest themselves as elemental totals far from 100%. We will return to the subject of normalization in the next chapter.

Figure 8.2 shows a flow chart for a typical Z-A-F calculation. We shall find that other steps may need to be added, having to do with modifying our intensity estimates as a function of composition and dealing with the need to compute the pure element standards from complex standards, or in the abscence of standards. However, the basic Z-A-F iteration loop will lie at the heart of most of the quantitative correction programs you will encounter. The particular choice of equation for the various terms, from among those presented here or others, will vary somewhat but will generally not make too much difference in your final answers. The literature, particularly in the earlier days of electron probe microanalysis, abounds with equations for the atomic number and absorption effects, in particular, with some models useful but limited in scope, some quite inferior to the more common ones, and some potentially much better but too complex for the level of computerization then available. Complete and reasonably accurate Z-A-F programs, including the calculation of all the necessary terms, are included in most commercial systems, and may be

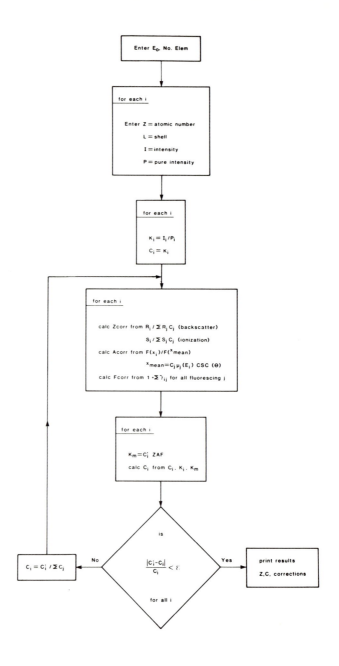

Figure 8.2 Representative flow chart for Z–A–F correction program.

obtained from noncommercial sources as well (such as the U. S. National Bureau of Standards).

The tables show two examples of the application of this calculation. The first is to a stainless steel, in which the elements Fe, Cr and Ni are close in atomic number (Z terms close to 1) but there are significant absorption and secondary fluorescence effects because the emission lines of Fe and Ni lie just above the absorption edge energies of Cr and Fe, respectively. The A term (which may be presented as the inverse of this value by some programs) shows the fraction of the generated radiation which is emitted, as compared to the pure element. If this number falls too far below 1 we would question the accuracy of the absorption correction model, as discussed in an earlier chapter. The F term is greater than 1 and shows the amount of extra emitted radiation from the element due to secondary excitation by X-rays from other elements (18% in the case of Cr). In the second example, the atomic numbers of Cd and S are farther apart, and the Z term for S shows the relative change in excitation in the compound as compared to the pure element. The radiation from both elements is somewhat absorbed in the sample, which we would expect for comparatively low energy X-rays such as the sulfur K peak especially. Fluorescence of S by Cd is very minor, and of course the Cd is not fluoresced at all since continuum fluorescence is ignored in this program.

Table 8.1 Z-A-F results for Fe -Cr -Ni stainless steel
        (22.5 keV., 23 degree takeoff angle)

	Cr K	Fe K	Ni K
weight %	18.70	72.24	7.51
intens.(cps)	224.28	582.56	40.28
pure intens.	1055.93	850.83	645.51
K-ratio	0.2124	0.6847	0.0624
Z-corr.	0.9801	1.0016	1.0379
A-corr.	0.9805	0.9366	0.8005
F-corr.	1.1820	1.0103	1.0000

Table 8.2 Z-A-F results for cadmium sulfide
        (12 keV., 30 degree takeoff angle)

	Cd L	S K
weight %	77.8	22.2
intens.(cps)	203.31	407.58
pure intens.	268.40	2210.30
K-ratio	0.7575	0.1844
Z-corr.	1.0097	0.9603
A-corr.	0.9643	0.8502
F-corr.	1.0000	1.0174

The particular form chosen for the Z-A-F correction is convenient because of the multiplicative factors, but not really correct of course. The intensity that is measured is actually the sum of the primary generated intensity, corrected for "stopping power" and "backscattering" effects, and the additional intensity produced by fluorescence by other elements and the continuum. Writing the terms in this way complicates the iteration process, and is only used in the rather large programs that also take continuum fluorescence into account.

Completely different approaches have been used with good results in particular cases, but seem unlikely to replace the Z-A-F method for general purpose work. First, of course, is the Monte-Carlo approach already described. The "Z" process is modelled directly by the summation of the ionizations produced by the electrons whose histories are simulated. Applying an absorption correction to the generated X-rays, and for the absorption events calculating the probability of secondary fluorescence, allows this to easily become a complete model of electron excitation. However, as stated before, the method produces intensities as a function of concentration, and is very difficult to adapt to the opposite problem of getting concentrations from intensities. The computer power needed to carry out the simulation with a large enough number of electrons to get good statistics is expensive, at least too expensive to consider using an iterative Monte-Carlo method to obtain quantitative results in routine analytical situations (and indeed the Z-A-F approach has proven itself able to give answers limited primarily by other sources of errors, such as counting statistics, in the majority of 'typical' cases).

Another approach, already mentioned, has been to use the equations presented earlier to predict the shape of the $\phi(\rho z)$ curve. From this, it is again easy (at least in the case of normal beam incidence) to incorporate absorption and fluorescence corrections by numerical integration. The total computational load is somewhat greater than for the ZAF method, but vastly less than Monte-Carlo. This method is particularly attractive for situations where the absorption correction is rather too large for the $F(X)$ equations to give good accuracy, while from a predicted $\phi(\rho z)$ curve the numerically integrated intensity emitted after considering matrix absorption is exact. This method has consequently shown itself particularly well suited to the quantitative analysis of very light elements whose X-rays are strongly absorbed (Brown, 1979b).

In the other direction, there are some applications for which Z-A-F calculations seem too slow, and for which simpler techniques are desired. At one extreme we could imagine simple calibration curves relating intensity to composition. This approach is rarely if ever used in electron excited microanalysis because of the difficulty of obtaining suitable standards whose concentration is known on a submicron scale, and because the multi-element nature of most samples makes the graphical interpretation of calibration curves very difficult. A simple extension of the calibration curve approach has found an area of successful application, however. The technique is known as the Bence-Albee (Bence, 1968) method, and is particularly well suited to the analysis of specimens containing very many elements (which would slow down a Z-A-F program, where internal loops covering all the elements' effects on each other would increase the computing time in proportion to nearly the square of the number of elements), but which vary over narrow concentration ranges and for which similar complex standards are available. Routine analysis of oxide minerals fits this description well, and geologists in particular are the greatest users of the method.

With this method, the interelement effects (combining Z, A, and F) are expressed as a matrix of coefficients that describe how the intensity of each element is affected (raised or lowered) by each other element. Over a limited concentration range, it is possible to express the total effect of a complex matrix as a linear sum of these coefficients times the weight fraction of each. With this scheme, very rapid convergence is possible since only linear sums are required. The interelement coefficients can actually be computed using a Z-A-F procedure for a particular set of elements and nominal concentrations. A second set of measured factors, which express the relative intensity for each element (equivalent to the pure element intensities) must be measured using one or more standard, compound samples on the particular instrument to be used.

The matrix of binary coefficients a_{ij} correspond to the factors in the expression

$$C = K a / (1 + k (a-1))$$

introduced previously. They are combined to form a factor β for each element

$$\beta_n = \sum_j C_j a_{nj} / \sum_j C_j$$

where the C's are elemental concentrations (weight fractions). As usual, these are first approximated by the K ratio (intensity ratio to the pure element) and iteratively improved until no further change takes place. In geological work, in particular, where standards somewhat similar to the unknowns are available (particularly with similar atomic number matrices) and secondary fluorescence effects are not too large, convergence is rapid and results are quite satisfactory even when very many elements are simultaneously analyzed. Since most of the use of this approach has been for minerals, the programs also commonly include oxygen, either by direct analysis or by its atomic ratio to each cation, and report additional information about oxide stoichiometry, possible chemical formulae, and so on.

In fact, many of the quantitative programs available report not only the elemental weight fractions they calculate, but also atomic or oxide percent, and the Z, A, and F factors so that the user can evaluate them for reasonableness. Warnings when the values exceed the range of best accuracy, and graphics displays of the results all help to make the programs "friendly" to the user.

Chapter 9

Standards:

Real, Complex and Imaginary

    In the preceding chapter the need for the pure element intensity as a standard to permit quantitative analysis was mentioned. Since it is often very difficult to obtain multi-element standards (samples whose composition is independently known) that are homogeneous on the fine scale needed for electron excited microanalysis, there are in fact many advantages to using pure elements as standards. For about half the periodic table, they are conveniently available as metal fragments that can be mounted, polished and analyzed repeatedly without damage. Sometimes it is useful to mount them semi-permanently on the microscope stage, and in other situations it is practical to mount the necessary standards with the unknowns, to be sure that they have a common surface orientation.

    For the k ratio (elemental intensity in the unknown divided by the pure intensity) to have meaning, the two measurements must of course be made under the same conditions. We would include in this category the accelerating voltage of the electrons, the surface orientation, and the beam current and counting time. If the intensities are reported as counts per second (with dead time correction appropriate to the type of counter and electronics used), the time constraint disappears. Similarly, since X-ray production is linearly proportional to the number of electrons that strike the sample, measuring the beam current and expressing the intensities as counts per second per unit of beam current (eg. nanoamperes) allows different currents to be used for standard and unknown. It happens that most electron microscopes do not have current meters that can read these values, and many that do are not set up to read the total beam current. The specimen current is the current that flows from the sample to ground. It differs from the total beam current because of the backscattering of electrons from the sample and also because of secondary electron production (which varies both with composition and geometry of the immediate surface layer of the sample). Consequently, in order to read total beam current it is necessary to construct a Faraday cup to collect all the electrons. The sketch in Figure 9.1 shows a simple design that can often be constructed by the user; it is simply a drilled hole in a low atomic number material such as graphite (although even aluminum will do). The hole is many times (at least 5) deeper than its diameter, and the backscatter coefficient is low. Consequently the fraction of backscattered electrons that escape is negligible (most are reabsorbed when they strike the walls of the hole). Also, the depth of the hole prevents the collection of secondary electrons by the field of the detector. A covering aperture will further help to contain the electrons. A simple test for this kind of collector is to place the beam within the hole and see if any signals can be detected by the various electron or X-ray detectors attached to the microscope. If not, then the total beam current is being collected and the reading from the "specimen current meter" represents the total beam current. It is usually wise to read this value before and after each analysis of specimen and unknown, which is facilitated by having the Faraday cup permanently attached to (or drilled into) the stage of the instrument. If the stage is motorized, the frequent re-checking of beam current (which in that case is also likely to be able to be read directly by the controlling computer) is very easy.

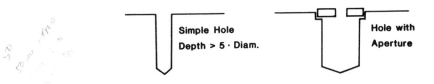

Figure 9.1 A simple Faraday cup to measure total beam current.

This re-checking of beam current is important because the stability of most SEM's and some microprobes is poor. The slightest change in the position of the filament in the electron gun, especially common when the filament is new and being annealed as it is heated, or old and close to failure, will alter the emission and hence the current. Some scan circuits also vary the beam current as a function of position, so it is wise to keep the specimen (standard, unknown, Faraday cup) at the center of the field of view during analysis. For some diffractive (wavelength dispersive) spectrometers, this is also important to preserve the proper alignment of the X-ray optics. Rechecking the beam current frequently enables questionable measurements to be repeated, or at least the current vs. time curve to be used to interpolate the likely value that occurred when the actual X-ray analysis was performed. With a wavelength dispersive system, this value may be different when each element is sequentially selected, and this must be taken into account.

If the SEM or microprobe (the case of the STEM comes up later, when thin sections are discussed) has no current meter, it is possible to use the X-ray spectrometer itself as a secondary reference. By mounting a standard pure element instead of a Faraday cup, and recording the intensity from it (whether or not it is an element that is present in the sample and needed as a standard is immaterial), relative changes in beam current can be monitored. Care must be taken to properly dead time correct these data. Then the changes in the reference standard intensity can be used to renormalize the intensities measured from standards and unknowns. Some automated systems include this capability as an alternative to a Faraday cup and current measurement, but the latter is to be preferred because the comparatively low yield of X-rays per electron makes the statistics of the X-ray signal poorer, and because the cleanliness of the reference standard (including surface contamination) or slight errors in position can substantially alter the measurement data.

Using some kind of reference for beam current variations, and accurate correction of dead time (and good preset time counting), the only "conditions" which must still be held constant between standard and unknown are the electron accelerating voltage and the surface orientation. We shall see that there are corrections that can be applied to adjust for changes in the latter term if the orientations are known, and previously discussed how the surface orientation can be measured. The voltage dependence is strong, and it is definitely necessary to match the values for unknown and standard. Fortunately, most electron column instruments are rather stable in gun voltage. Problems can still arise if the specimen surface charges. This may occur because of inadequate specimen conductivity or a poorly applied surface conductive layer, usually carbon rather than heavy metals when microanalysis is carried out, to reduce the absorption of X-rays (the exceptions to this general practice will be mentioned when ultra-low energy analysis is considered). In theory, specimen charging does not preclude accurate analysis, provided that the voltage of the electrons as they strike the surface on standard and unknown can be matched. Unfortunately, this is rarely practical because the charging fluctuates with time and the voltage drop experienced by the electrons is not controllable. This voltage may be quite large -as much as 10 to 15 keV. -without making it impossible to get normal electron images.

Still, we need to be able to accurately measure the electron voltage not only to monitor it for changes, but because it enters many of the correction terms that have been described. The voltage can rarely if ever be taken directly from the setting on the microscope control console, or even be read confidently from a meter provided there. This is due partly to the possibility of specimen charging, but more fundamentally to the inherent nature of the high voltage control circuits and the electron gun itself. Most manufacturers take no pains to adjust the actual high voltage to agree with knob settings or meter readings (which do not measure the high voltage itself, but a control voltage elsewhere in the circuit), because it is unimportant for the principle imaging role of the instrument, and because it would never be accurate enough for quantitative corrections anyway. The voltage is actually applied to the grid and anode of a typical electron gun, with the filament voltage controlled by a bias resistor so that the actual voltage depends on the current flowing in the gun. Changing bias setting (selecting a different resistor, aligning the gun, or changing the filament) will inevitably change the actual accelerating voltage.

The practical solution is to depend on the instrument settings for approximate voltages, but to measure the actual voltage with the X-ray spectrometer itself. One technique that can be used is to plot the intensity for a characteristic line of one or more elements whose absorption edge energies cover the range of interest, as the voltage setting is changed. Since the intensity curve will fall to zero when the accelerating voltage equals the absorption edge, it is possible to calibrate an instrument in this way. A far simpler method (Solosky, 1972) is to determine the accelerating voltage by the Duane-Hunt limit: the energy (or its equivalent wavelength) at which the continuum ends. The shape of the background is nearly linear with energy in this range, and either by collecting an ED spectrum or scanning a WD crystal, the cutoff point can be determined. As shown in Figure 9.2, this can be complicated by several factors. The background intensity is low near the cutoff point, so that the counting statistics are poor and long times are needed (and even then the scatter in the data makes it necessary to fit the line through many points). Also, the ED detector suffers from pulse pileup at high count rates (which we would like to use for this purpose because of the statistics problem), and rather than dropping to zero, the measured continuum intensity may instead be superimposed on a more-or-less flat shelf of pile-up counts. Extrapolating the falling continuum curve to the intersection with the flat shelf gives the cutoff voltage, with the expected problems of fitting lines to low count data. It is also clearly essential that the system be well calibrated. For either an ED or WD system, the area of most frequent use is likely to be rather far from that at which the continuum cutoff occurs, so that calibration based on the peak position of a few common characteristic emission lines of elements may not extrapolate well to the cutoff point. For ED systems, it may even be necessary to use a different amplifier or ADC conversion gain setting to record the cutoff, and in that case, careful calibration of the energy scale being used is essential. However, if reasonable care is taken in all of these areas, determination of the incident electron energy to within a few hundred volts is routinely possible, and this is adequate for quantitatively accurate corrections to be computed.

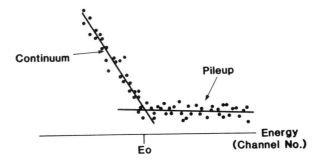

Figure 9.2 The continuum cutoff point identifies the electron accelerating voltage.

Compound Standards

For the half of the periodic table which can be obtained in the form of pure metals, it is possible to directly measure the pure element intensity. For many other elements, some of great practical importance, the element is either non-conducting, unstable in vacuum or under the electron beam, or not even a solid at room temperature. In this case we cannot measure the pure element intensity directly, and so we make use of another approach. Many elements are readily available in compounds whose stoichiometry is known (for instance, consider the difficulty of measuring the pure intensity of chlorine, and the availability of NaCl). By choosing compounds with well defined elemental composition and fairly large amounts of the elements we need, we can measure intensities and then compute what the pure element intensities would be. The earlier equation relating concentration to k ratio was:

$$C = k / (Z\ A\ F)$$

where $k = I_{unk} / I_{pure}$

If we introduce the symbol P_i for the pure intensity of element "i", the equation can be re-written as:

$$P = (I/C) / (Z\ A\ F)$$

Where I is the intensity measured on a sample containing concentration (weight fraction) C. The Z, A, F factors are still functions of the concentrations of all the elements in the sample, and are calculated using the same equations discussed previously. However, since the concentrations are known (this is a standard, after all), no iteration is involved. Most programs that can perform Z-A-F calculations on unknowns can also run the calculation "backwards" to get pure intensities from complex standards (sometimes this process is hidden from the user, with the full set of data of standards and unknowns entered and only the final answer returned). These standards may be convenient simple stoichiometric compounds or materials much like the unknowns which are to be run; the latter approach is particularly common in the analysis of minerals, where a fairly large number of well-analyzed known minerals are available and can be used as standards. Sometimes the use of complex standards also saves time, particularly when ED systems are used, since one or a few standards may contain all of the elements present in the unknowns. In the limiting case, a different complex standard may be analyzed to determine the pure intensity for each element in the unknown. When selecting complex materials for use as standards, remember that they must be homogeneous on a submicron scale, that all of the elements present must have known concentrations, and that compounds containing a high concentration of the element are preferred for use as a standard, compared to ones were the concentration is low. Also, the magnitude of expected errors in computing backwards to get pure intensites is the same as in the normal calculation for unknowns, so that very high values of χ (due to high matrix absorption) or the use of an inappropriate accelerating voltage are still to be avoided.

Interpolating

Even with the ability to use compounds or complex materials as standards, and back-calculate the pure element intensity from them, there are times when no proper standard is readily available. In those cases, we may have either a pure or compound standard for other nearby elements in the periodic table, but not for the element we need. If the pure intensities (measured or calculated) for a series of elements are plotted against atomic number, we generally find a smooth curve can be drawn through the points (Blum, 1973). This encourages us to try to interpolate with such a curve to find the values needed for elements not directly measured. For instance, Figure 9.3 shows a curve plotted through the measured intensities (at a 25 keV accelerating voltage and 35 degree takeoff angle) for several transition metals. The missing point for manganese could be confidently interpolated from such a plot. For the ED spectrometer, this technique has been carried to the point of programs that fit curves (using quasi-theoretical functional forms with plenty of adjustable constants) to a series of measured values, and provide pure intensity values for all of the intervening elements. The curve is quite regular and well behaved, for reasons we shall see, at least as long as we are looking at all K, L or M lines and not combining measurements from different shells. It is even possible in some cases to use curves measured with one detector to standardize another.

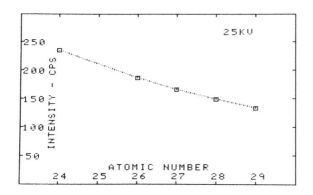

Figure 9.3 Plot of measured pure element intensity versus atomic number.

For the WD spectrometer there is less universality in the curves, and more discontinuities are likely to occur because the efficiency of the spectrometer is a strong function of angle, crystal, slit setting and other aspects of the focussing geometry. Two crystals of the same nominal type will rarely reproduce each other's intensity, and even realignment of a spectrometer may not only change the absolute magnitude of the cure, but its shape as well. Nevertheless, with either ED or WD spectrometers, the measurement of standard intensities from elements that bracket one that is needed and unavailable is often a very satisfactory way to obtain the necessary data. Extrapolation is more dangerous. Unless the shape of the curve is very well known (which is more practical with ED systems where we shall shortly be calculating it from first principles), going more than one or two elements beyond the end of the region over which a curve was fitted is usually an invitation to surprise if not disaster.

105

Calculating Pure Intensities

In principle, we should be able to compute the absolute intensity from a pure element from the fundamental processes of electron penetration and X-ray absorption. The overall relationship is just (Russ, 1974b):

$$P = R\,W\,\omega\,L\,F\,T$$

where R is the probability that the electron deposits its energy in the sample (our old friend the effective current or backscattering correction), W is the total ionization probability which we shall get from the ionization cross section and stopping power, ω is the probability that the ionized atom emits an X-ray (the fluorescence yield), L is the probability that the X-ray is in the particular line (e.g. K-α) that we are going to measure, F is the probability that the X-ray travels out of the specimen without being absorbed, and T is the probability that the X-ray is detected (the spectrometer efficiency). We shall examine each of these is turn, realizing that most are calculable using equations we have already encountered.

The effective current factor or backscattering correction R is identical to that described in a previous chapter. For the particular case of a pure element, we know the atomic number Z, the absorption edge energy E_c, and of course the electron accelerating voltage E_o. Similarly, we know how to calculate fluorescence yield ω, and the absorption correction $F(\chi)$ where χ is the product of matrix mass absorption coefficient (here simply the element's absorption coefficient for its own X-rays) with absorption path length (dependent on surface orientation). It is sometimes convenient to calculate the elemental self-absorption coefficient without resorting to the complete model for the mass absorption coefficient for any energy X-ray in any element; the equation below can be used for the principal (K-α, L-α-1, M-α) lines.

$$\mu\,(\text{self}) = A\,Z^{B}$$

shell	A	B
K	2.94477×10^{-5}	-2.56464
L	1.90666×10^{-7}	-2.72348
M	7.22867×10^{-9}	-3.58163

The probability of an atom emitting an X-ray in the line we are measuring is simply the relative line intensity. These can be found in tables, or approximated from simple equations fit to empirical data such as those shown in a subsequent chapter dealing with stripping of peak overlaps. In some cases it is necessary to use more elaborate calculations which take into account the effect of electron voltage. This is particularly true in dealing with the L lines where the presence of three subshells with different excitation energies substantially complicates the relationships.

The only remaining terms are the spectrometer efficiency T and the total X-ray yield W, which we shall now take up individually. From the ionization cross section Q, which is a function of electron energy, we can get the production of ionizations n by integrating along the path length s (the distance along the actual zig-zag path and not the depth in the sample) the relationship:

$$dn/ds = Q \ N_{Av} \ \rho \ / \ A$$

where Avagadro's number, atomic weight and density define the number of atoms encountered. The cross section for electrons is (Powell, 1976):

$$Q = 2 \pi \ b \ e^4 \ / \ (E \ E_o) \ \log \ (4 \ E \ / \ B)$$

where the 'best' values of B and b for K shell excitation are $4 x E_c$ (taken as the limit of theoretical relationships as overvoltage approaches 1) and 0.606, respectively. Using these values, we can rewrite the expression in terms of the overvoltage $U = E_o \ /E_c$ as shown below, and plotted in Figure 9.4 as a universal curve of $Q \ E_c^2$ versus U:

$$Q \ E_c^2 = 7.92 \ 10^{-20} \ \log \ (U) \ / \ U$$

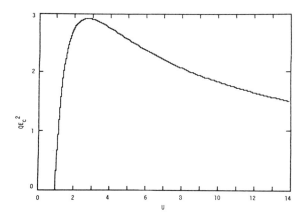

Figure 9.4 Plot of QE_c^2 versus overvoltage U.

Incidentally, the units for this graph are in $cm^2 \ keV^2$, and because of the very small constant value, this implies that the cross section for electron excitation in this energy range is very small (the process is unlikely to happen, and the electron may pass 'through' several hundred atoms before ionizing one). This means that our rather simplified Monte-Carlo model in which the electron proceeds by relatively long (by atomic dimensions) straight segments is acceptable.

Integration of the cross section can be accomplished by using the Thomson-Widdington energy loss relationship:

$$dE/dx = -c\rho / 2E$$

where ρ and E are density and electron energy, and c is a "constant" that actually varies from about 1×10^{-5} for 1 keV electrons to about 4×10^{-5} at 40 keV. Integrating the cross section gives the total number of ionizations (of the K shell) as a function of overvoltage:

$$n_K = 9.535 \ 10^4 / (A \ c) \ (U \log(U) - U + 1)$$

A complete treatment of the process of generation of X-rays from a pure element should not neglect the fluorescence of the element by Bremsstrahlung X-rays, however. This adds an additional

$$1.46 \ 10^{-8} \ Z \ (Z-2)^2 \ (U \log(U) - U + 1)$$

from which we can reconfirm our earlier remarks about the relative magnitude of continuum fluorescence to primary generated intensity as a function of atomic number of the matrix. The ratio goes approximately as:

$$3.24 \ 10^{-13} \ Z^4 \ c/R$$

where R is the correction for backscattering. This fraction varies from less than 0.02% for aluminum to about 7% for copper, and to 44% for silver (of course, using a high enough beam voltage to excite silver K α-rays is unusual in SEM work, and would produce much more continuum at high energy than the use of lower voltages to excite the L lines).

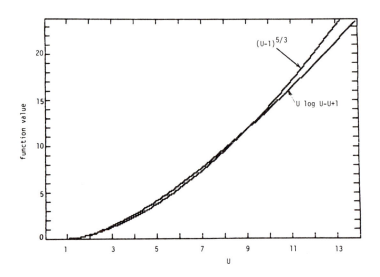

Figure 9.5 Comparion of expressions for overvoltage dependence.

Green and Cosslett (1961) substituted the expression:

$$0.365 \, (U-1)^{1.67}$$

for $(U\log(U)-U+1)$, as being easier to calculate. Experiments have confirmed that for a variety of elements, the dependence of intensity on overvoltage follows $(U-1)^n$ where the exponent lies between 1.63 and 1.67. Figure 9.5 shows the agreement between the two expressions, which diverge only for rather high overvoltages. The dependence upon $(U-1)$ is very logical, since it is when U equals 1 that the incident electron has just enough energy to produce an ionization, and as U exceeds 1 we would expect the probability of excitation and the total number of such ionizations to increase. The roughly constant exponent has been confirmed by experiments over a wide range of U values, and for K, L, and M lines.

Combining these expressions, we find for the emitted (not detected) K intensity from a given element, we have:

$$\omega L \, F(\chi) \, (2.77 \times 10^3 \, R/Ac + 4.24 \times 10^{-10} \, z(z-2)^2) \, (U-1)^{1.67}$$

For the L shell the result is similar, except for the constants. In the equation below the fluorescence yield ω is for the L III subshell and the factor J is the relative absorption edge jump ratio for the L III subshell as a fraction of the total L absorption edge.

$$\omega L \, F(\chi) \, (3.83 \times 10^3 \, R/Ac + 5.90 \times 10^{-11} \, JZ(z-10)^2) \, (U-1)^{1.67}$$

The curves in Figures 9.6 and 9.7 plot these functions in two ways: first for different elements at a constant overvoltage ratio, from which we can see the fundamental ability of different elements to generate X-rays, and then at a constant accelerating voltage, showing the relative intensities that will actually be emitted from a series

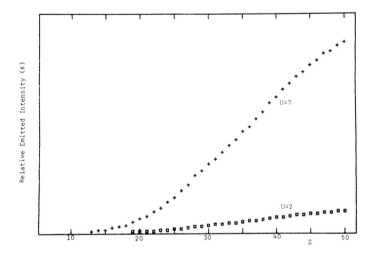

Figure 9.6 Emitted intensity as a function of atomic number, at constant overvoltage.

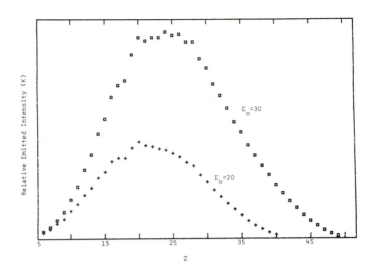

Figure 9.7 Emitted intensity as a function of atomic number, at constant accelerating voltage.

of pure elements under fixed excitation conditions. The absorption correction was calculated based on perpendicular electron beam incidence and a 30 degree takeoff angle, and a mean value of 2.5×10^{-5} was used for c. All of the other parameters were calculated using equations already presented.

Spectrometer Efficiency

Spectometer efficiency is defined properly as the fraction of all X-rays (emitted isotropically in all directions) that are actually counted. This depends on the solid angle intercepted by the spectrometer (the slits and crystal for a WD system, the detector itself for an ED system), on the transmission characteristics of any windows in the X-ray path, the efficiency of diffraction for a crystal, and finally the detection efficiency of the detector itself (the Si(Li) diode or the gas-filled proportional counter). Although the emitted intensities calculated using the expressions above are on a per-unit-solid angle basis, since the spectrometer solid angle is usually a constant for a given system we will not include it here. For a WD system the diffraction efficiency depends so strongly on the individual crystal used, and its degree of perfection, that we shall not attempt to calculate overall system efficiencies for these, but shall restrict our further efforts to the ED spectrometer.

For a Si(Li) detector, the total detection efficiency is the product of several terms: 1) the probability that the X-ray gets through the entrance window(s) represented by the beryllium window, gold contact layer, and "dead" silicon layer; 2) the probability that the electron does not penetrate completely through the detector and go out the back; and 3) the probability that the absorption of the X-ray in the detector is by the photoelectric effect and not due to scattering. For the energy range we normally deal with, up to about 20-40 keV., the total scattering cross section in Si never exceeds a few percent of the photoelectric cross section. Consequently, it is often ignored. The result is that based on the window transmission and total detector absorption, it is common to hear of detector efficiencies "approaching 100%" when in fact it would be more proper to take both the solid angle and fractional photoelectric cross section into account.

The absorption of X-rays in the detector itself and in the various layers of the entrance window can be calculated simply from the Beer exponential law. If the thickness of each layer is known, the spectrometer efficiency becomes:

$$T = e^{(-\mu\rho\, t_{Be} - \mu\rho t_{Au}\, \mu\rho t_{Si})}\, (1 - e^{-\mu\rho t_{Det}})$$

where the subscripts identify the beryllium window, gold contact layer, silicon dead layer and total detector thickness for which the thickness, density and mass absorption coefficient (a function of X-ray energy) are to be evaluated. Only the nominal value of each of these thicknesses is normally known. The beryllium window is a thin, rolled and somtimes chemically etched sheet of metal only a few grains thick, and its surfaces are far from planar. The thickness and density may vary considerably from the manufacturer's specified value (which properly is only an upper limit to thickness anyway). The gold contact layer is often thin enough to ignore, but it too is not a smooth continuous film, but in the thickness range of about 100 angstroms is more probably a filamentary structure with thick islands and barren areas between. The thickness of the dead silicon layer depends somewhat on the applied bias voltage, and is apt to change (for the worse) if the detector is cycled to or stored at room temperature, due to the increased mobility of the lithium ions. Even the total detector thickness cannot be taken simply as the physical thickness of the device, because the rear contact area normally consists of a relatively thick layer of strongly doped material which is partially removed by etching. The active or intrinsic region is thinner than the total device, by varying amounts depending on the fabrication.

Because these parameters vary from device to device, and sometimes as a function of time, they should be measured for each spectrometer. Even the beryllium window thickness may appear to increase if oil vapors deposit on the front face (it is usually a few degrees cooler than the surroundings because of the proximity of the cooled detector behind it) and increase the absorption of X-rays. By measuring the intensity of X-rays at a series of energies where absorption in the window or transmission through the detector is significant, and for which the mass absorption coefficients can be calculated, the various (ρt) mass thicknesses can be determined from a set of simultaneous equations. This requires knowing the intensity of the radiation arriving at the detector, and therein lies the rub. At high energies (from 20-30 keV.) it is possible to use K lines from elements in the atomic number range from 40 to 50, provided that the electron beam instrument has a high enough accelerating voltage to excite them (if it does not, then no analysis will be performed up there anyway and the high energy efficiency of the spectrometer is moot). The relative intensities of these lines can be estimated quite well using the fundamental equations presented above. At the low energy end, from about 2 keV down, the pure element K line intensites are strongly affected by absorption within the sample itself, to the point where $F(X)$ corrections may have doubtful accuracy. Also, electronic limitations in the spectrometer may influence the measurements; for instance, some pulse pileup rejectors will cut off low energy pulses which cannot be distinguished from electronic background noise, giving the appearance of additional physical absorption of the X-rays. Finally, measuring a series of lines from different standard elements is time consuming and prone to errors if the instrument stability is imperfect.

One of the two most common approaches to characterizing the low energy efficiency of ED detectors is to measure the continuum generated from a known sample, for instance a piece of graphite (carbon, Z=6), and using the known expected shape of this curve (by fitting at higher energies where window absorption is neligible), compute the absorption at lower energies due to the window (Fiori, 1976a; Russ, 1977a). This in principle can give separate values for the thickness of each absorbing layer, for later use in computing the spectrometer efficiency in other calculations with standards and unknowns. If there are errors in the values because of wrong assumptions about mass absorption coefficients (which are doubtful in this energy range) or even the composition of the layers (for instance treating an oil film as extra beryllium), they will tend to be self cancelling when the same parameters are applied later on.

An even simpler method is to measure the ratio of L to K lines fo some convenient element such as Ni, Cu or Zn (which have L lines at or below 1 keV. where window absorption is significant but not total). Using the equations already given, we can calculate what the emitted L/K ratio should be (including absorption in the sample), and from that, what the additional window absorption of the L lines is. The K lines are in the energy range where window absorption and through penetration are both negligible, and the detector is maximally (though not really 100%) efficient. Since this gives only a single equation for the absorption, it is not possiable to separately compute the thickness of each layer. Instead, a generalized function for the low energy transmission ("detector efficiency") of the entrance window is used that contains a single ajustable parameter, whose value is obtained from the single measurement of the L/K ratio. This is less accurate than the more detailed calculation at low energies, because it neglects the discontinuities in transmission efficiency which correspond to the absorption edges of the silicon and gold (as shown in Figure 9.8), but again the reapplication of the same factor with standards and unknowns reduces the impact of the error, and in any case this is a sufficiently simple procedure to permit frequent and easy checking of the detector to monitor changes in light energy performance which could signal oil contamination or mis-adjustment of the pileup rejection discriminator. The functional form is (Russ 1974b):

$$T = e^{-WTC/E^{2.8}}$$

where WTC is the "window thickness constant," about 1 for a typical 7 micrometer thick Be window and Si (Li) ED spectrometer, and the exponent of 2.8 on the energy E of the X-ray describes the general variation of mass absorption coefficient with energy. Figure 9.8 compares these two transmission efficiency expressions for a typical spectrometer system.

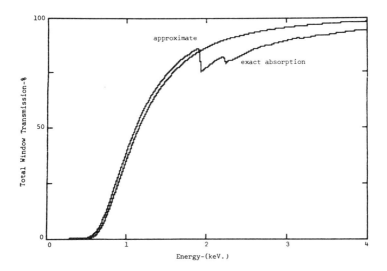

Figure 9.8 Models for window absorption contribution to detector efficiency.

Using Theoretical Intensities

The equations for absolute intensity for a given element are practically never used as standards in quantitative electron excited microanalysis, because they simply are not accurate enough. The constants that appear in the expressions, and the fact that some terms such as exact solid angle have been neglected, mean that these expressions should give the proper relative dependence of pure element intensity with voltage, atomic number, specimen geometry and so on, but can't be trusted for the absolute values (which would also require absolute measurement of the beam current). Instead, if we casually throw out the constants and the continuum fluorescence dependence on Z, and retain just the functional form of the exressions, we can use them for the fitting and interpolating process mentioned before, Within a give shell, the pure element intensity should vary as:

$$R \omega \ L \ F(X) \ T \ (U\log(U)-U+1) \ / \ A$$

where the term involving U can be replaced by $(U-1)^{5/3}$ if desired. All of the other terms are readily calculable as functions of Z, E, and specimen geometry (except for L, the relative line intensity, which we can make equal to one if we use the total shell intensity measured with an ED spectrometer). By linear least-squares fitting we can determine the constant of proportionality for this relationship from a number of pure element intensity values determined on a particular system, using either pure or compound standards. Having the proper functional dependence on Z makes it more reasonable to interpolate over greater ranges of Z, or to extrapolate where necessary. Plots of measured data against this form of curve show quite good agreement. In principle it is even possible to measure only a single pure intensity (remember the single element reference standard we substituted for a Faraday cup before) and from that determine the entire pure element intensity curve.

Within a single shell, particularly for K shell elements, this works quite well indeed. When the ratio of lines between shells is involved, there is an additional factor that must be introduced into the equation. From the fundamental absolute expressions above, and neglecting continuum fluorescence contributions to intensity, the L to K ratio should be 1.38:1 (but the value of c is not really constant and higher energies are involved in K line excitation). On another basis, considering just the relative number of electrons per shell, the M:L:K ratio should be 9:4:1, but this neglects screening effects. Empirical measurements give M:L:K factors of 4.36:3.56:1, but the possible Z and E dependence of these factors has only been lightly examined.

This means that within a given shell the use of a simple model for relative elemental pure intensities is rather good, even for quantitative work, but that between shells it is dangerous, and we would like to have at least one point to determine the section of the curve for each shell of interest. One very practical way to do this is to mount, more or less permanently on the microscope stage, a few elements such as Si, Ni, Mo and W which cover the periodic table and allow the pure element curves to be established whenever needed, for a given accelerating voltage and orientation. These particular elements have the virtue of being easy to polish and keep free from oxidation (compared, say, to aluminium and iron), but any reasonable set of elements will do.

Of course, the curve will still correspond to the orientation of the surface of the standards (most likely the same as the stage), and not to the unknown. But this can be dealt with using the functional form for pure element intensities. The R (backscattering correction) and F(x) (absorption) terms include specific corrections for the inclination of the specimen surface and the X-ray takeoff angle, and consequently it is straightforward to adjust an element's pure intensity value from one orientation to another. We must respect the limited accuracy of the approximations used, of course, and not expect to get good results at very high tilt angles or very low takeoff angles, but these conditions do not lend themselves to good quantitative results anyway. With this tool, we can measure standard intensities at one orientation and convert the pure intensity values to a different orientation to match the surface inclination of an unknown (we still have to determine that orientation, of course). In principle, the same adjustment should be possible as a function of accelerating voltage, but because of the strong dependence on voltage and the approximate nature of the expressions, this is definitely not recommended, and standards should be measured at the same voltage as unknowns.

No Standards at All

Given that we can compute the relative shape of the pure element intensity curve, there is another way to get k-ratios (the starting point for Z-A-F analysis) that requres no standards measurements at all (Russ, 1978). Instead we depend on the fact that the elements we have analyzed must add up to 100%. This may include an un-analyzed element such as oxygen, if its ratio to each of the analyzed elements can be determined independently (ie. from the user's knowledge). The concentration of each element is first set equal to I/P where P is the relative pure element intensity, and then the k-ratios are normalized to add up to 100%

$$k'_i = k_i \, / \, \Sigma \; (k_j)$$

Using this as a first approximation for concentration, the Z-A-F scheme is carried out. At each iteration, the k ratios are renormalized to force the calculated concentrations to total 100% (including the extra element defined by its ratios to the analyzed elements), by calculating

$$k''_i = k'_i \, / \, \Sigma \; (C_j)$$

which is much less dangerous than simply normalizing the concentrations to total 100% at each iteration, but will still hide from the user any indication of problems that occur if an element is present that was not anticipated or included in the analysis of minor amounts of heavy elements in a basically organic matrix, since the analyzed elements will take on a total of 100% and hence vastly exaggerate the magnitude of their interelement effects (the size of the Z, A, F terms).

Although this technique is very fast, since no standards at all have to be measured, and consequently very easy (since the user can push a button and needs to know little or nothing about what is going on), it is obviously easy to get into trouble by using it in the wrong way, and because the answers are printed out with the usual number of decimal places by the computer, to assume that the accuracy of the result is reflected by the numbers. When different shells are involved in a single analysis, or when light elements (for which absorption corrections are large) are included, or when the analysis conditions are generally those that may be expected to produce poor accuracy in the corrections, the use of normalization with no standards will not solve the problem.

On the other hand, it is a legitimate method for rapid semiquantitative and sometimes quantitative analysis. There are even circumstances in which it will do better than the use of conventional standards, such as in a microscope with an unstable beam current. The spectrum integration by the ED spectrometer gives intensities that are quite correct when ratioed only to each other, but could not be accurately ratioed to pure element intensities derived from standards because the beam current variation would introduce a larger error than that from calculating the entire pure element intensity curve.

Chapter 10

Accuracy and Errors

Thus far, the various interaction processes and correction equations presented have shown differences from each other and from the "exact" truth which have been described frequently as "negligible" or "acceptable". It is now time to consider the overall accuracy and the limiting errors that come about in the final results. Most analysts wish to characterize their quantitative answers in terms of expected overall accuracy, or to take steps to improve it by attacking the greatest source of error. Since there are many independent sources of error each with its own magnitude, we can write the following equation to relate the square of the total standard deviation to those of each contributing factor:

$$s^2_{total} = s^2_{stat} + s^2_{sp.proc.} + s^2_{model} + s^2_{std.} + s^2_{cond.} + \ldots$$

Each of these contributions will be discussed below. The three dots at the end of the equation indicate that there are indeed other sources of error that we will not consider. They may sometimes be so major as to wipe out any attempt at useful results, but depend more on gross malfunction of the equipment or of the human working it. For instance, if the pulse pileup rejector is malfunctioning or its discriminator maladjusted, the count rates at low energy can be hopelessly distorted.

And obviously the standards must really be what they are assumed to be. In my own personal experience, I recall an instance when a cadmium standard was needed for quantitative analysis of some near-stoichiometric CdS. The cadmium L and sulfur K lines were used, so that standards were clearly called for. A piece of iron pyrite (FeS_2) was used for the sulfur standard, but I had a block of standard elements (mounted and polished) that contained cadmium as one element. The handy grid map accompanying it showed the Cd position in one corner of the grid, and at the first corner I looked at, I found a specimen whose peaks matched reasonably well with the Cd L spectrum. Had I looked really carefully, I would have seen discrepancies, but after all, I _knew_ this was a standards block and contained Cd. Of course, it wasn't; the opposite corner in the standards grid was uranium metal, whose M lines are not too far from the Cd L lines. The quantitative results were (fortunately) very far from anything reasonable, and so the error was eventually discovered. This kind of error, which can happen to any operator regardless of experience level if haste or Murphy's Law strikes, is not included in the equation above.

117

Counting Statistics

    Normally when we count X-rays, they obey the statistics of independent random events: Poisson statistics, which for large numbers are well approximated by Guassian probability distributions. Hence, for a number of X-rays N, the relative standard deviation should be given by $\sqrt{(N)}/N$. This suggests that if we have 10,000 counts for an element, the one-sigma expected error should be $\sqrt{(10000)}/10000 = 1\%$, and in 68% of the analyses, the data should lie within this range (95% of the time within 2%, etc.). Unfortunately, things are a bit more complicated by the fact that we are really counting several things at once, which cannot be distinguished as they are counted. At a given energy or wavelength setting of the spectrometer, the X-rays counted include in addition to the characteristic lines from the element of interest, the Bremsstrahlung X-rays at the same energy and any characteristic X-rays from another element that overlap the peak of interest. The net intensity for the peak of just the element of interest is determined by subtraction. For the moment we shall assume that the subtraction is done <u>perfectly</u>, that is that the fitting of background and peak overlaps is absolutely correct (those errors come next). Even so, the statistical uncertainties in those numbers exist, and increase the uncertainty of the net counts.

    As a good example, consider counting cars. By standing on a bridge over a busy freeway (perhaps a random process, at least at rush hour), we may count all of the cars for a given period of time. Then we will count those cars that are blue, and those that are red. If the only other possible color were green, then we might expect to be able to estimate the rate of green cars passing under the bridge by subtracting the rates of the blue and red cars from the total rate. However, because each of these numbers has a statistical uncertainty, the net value we obtain for green cars includes all of the counting uncertainties: for the total rate, and for the red and blue rates. If the green rate is very low, and we try to get it by subtracting one large number (red plus blue) from another large number (total), the net result will have a very large uncertainty that may make it impossible to tell if there are any green cars at all (the case of a trace X-ray peak at the detection limit).

    For the X-ray intensities we want, the standard deviation from these purely statistical considerations in the net counts for the element of interest is thus not $\sqrt{(N)}/N$, but rather:

$$s_N = \sqrt{(N + 2\ (BG + OV))}\ /\ N$$

where BG is the background intensity and OV is the sum of any overlaps present. Consider the case of an ED system used to analyse a stainless steel containing 1% manganese. Rather than 10000 counts, we may be lucky to collect 1000 for the Mn K-α (most of the counts recorded will obviously come from the major elements Fe, Cr and Ni). Based on $\sqrt{(N)}/N$ this would still lead us to predict 3% relative standard error for the manganese. But in this case, the background counts under the peak will probably contribute another 1000 counts, and the overlapping K-β peak from the larger concentration of chromium may contribute another 2500 counts. The relative standard deviation for manganese thus becomes $\sqrt{(8000)}/1000$ or about 9%. Figure 10.1 shows the one-sigma error expected for different intensities as a function of the peak to background ratio of the measurement (for this purpose including overlaps with background).

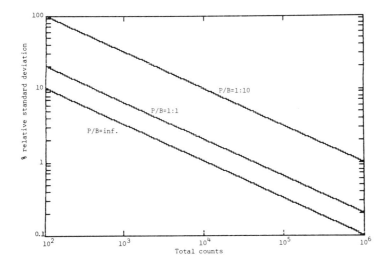

Figure 10.1 Statistical error in net X-ray counts as a function of the total counts and underlying background or overlaps.

For minor elements, or ones with strongly overlapped lines, the statistical errors can be quite large (greater than 10%). Counting for longer times improves the statistical accuracy, but only in proportion to the square root of time (and very long time analyses are not only impractical in many cases, but may cause other problems if the instrument or sample stability is poor, or surface contamination occurs). One of the greatest advantages of the WD spectrometer is its excellent peak resolution; the narrow peak widths cover proportionately less background, and often avoid any peak overlaps altogether. This improves the statistical accuracy, so that in fact it may approach $\sqrt{(N)}/N$. In addition, the WD systems can process much higher counting rates than ED spectrometers, especially for minor elements, due to the ability to reject X-rays from the major lines by diffraction, which would swamp the ED detector.

Spectrum Processing

The next chapter takes up methods of spectrum processing for ED spectra in more detail. It is nonetheless appropriate here to include a few general comments about accuracy. For WD spectra, since the peaks are so narrow and generally do not overlap each other, usually all that is necessary is to fit a straight line under the peak from background locations on either side. Sometimes the use of a single background measurement at a wavelength near the peak is all that is needed. It is important to be sure that the peak setting is really on the peak, because small changes in sample surface position can cause peak shifts in the spectrum even between standard and unknown, and longer term alignment and temperature effects can also introduce errors. With computer-controlled spectrometers, step counting across the peak and integration of the total intensity gives the most reliable estimate of intensity,and there are also fitting routines that locate peak centers automatically.

For energy dispersive spectra, the greater peak width covers enough background that the curvature of the continuum, particularly at low energy, cannot be ignored. Linear interpolation from one side to another will lead to quite erroneous results when curvature or discontinuities due to absorption edges are present. Fitting of more complex functions, either higher order polynomials or quasi-theoretical functions that model the shape of the background, can in principle correct for this provided that the functions are adjusted to fit the measured spectrum at locations that are truly background, free from peaks or other artefacts.

The deconvolution of peak overlaps is generally accomplished by least squares fitting of peak shapes which have either been mathematically generated from semi-theoretical equations, or measured and stored from standard samples containing large concentrations of the elements (and from which the background must also have been removed, in another process introducing some error). These fitting techniques lead to useful results in most cases, as attested to by the many published quantitative results using ED spectrometers. However, quite large errors can be produced in a few specific cases. One such problem comes from the use of peak shapes or ratios of peaks (major and minor lines) that do not match, either because of erroneous calculations (e.g. not taking into account the matrix absorption for each line separately, not calculating the different excitation of each L subshell as a function of electron voltage, not adjusting the equation used to generate the peak shape for energy-dependent tails as well as width, and so on). These can easily introduce errors of several percent. Consider as an example the case of manganese in the stainless steel, cited above. The chromium K-α peak will be perhaps 20 times larger than the manganese K-α , depending on concentration and to a lesser extent upon excitation conditions. The K-β /K-α ratio for chromium is about 1:8, from various tables and fitting functions. The difference between using 12% or 13% for the K-β peak height produces a 20% relative change in the size of the manganese peak left after the subtraction.

Even more important in many cases is the problem of spectrum calibration. Fitting programs usually assume that the peaks lie either at the channel positions expected on the basis of a nominal or predetermined calibration factor, say 20 electron volts per channel, or at the same energy calibration as used to record the standard spectra to be used for fitting. If this is not so, either because the original calibration was sloppy or because there has been a slight shift in position due to thermal or electrical ground shifts, or due

to count rate effects in the amplifier, the results of the fitting may
give very wrong intensities. Furthermore, there may be no indication
of a problem, since the agreement of the fitted data with the original
may be quite good (in other words, the peaks used to fit to the
measured spectrum can be added together, times their respective
fitting factors, to produce a result that matches very closely the
spectrum measured with the overlapped peaks in different positions and
with different relative intensities).

As an example of this, Figure 10.2 shows the variation in net
elemental intensities for Pb L-α and As K-α with calibration shifts.
The χ^2 factor is a measure of the degree of match between the
measured and fitted results, and while it is not particularly
sensitive to the shifts, it does show a minimum at the wrong place (it
is dominated by statistical effects). The difference between the fit
giving the best match to the measured spectrum and the "true" arsenic
and lead intensities is about 25% for the As and 32% for the Pb. While
this is a rather severe example, and many peak overlaps which are
farther apart (the Pb L-α and As K-α peaks are separated by only about
20 electron volts), the same type of error occurs even for single
isolated peaks if the calibration is off. Peak shifts of only a few
eV. are practically invisible to the human observer, and within the
normal operating specifications for most manufacturers, yet can
introduce errors of several percent in the intensity values.
Adjustment of spectrometer calibration to the necessary precision
requires computer assistance, which is now provided by many systems.

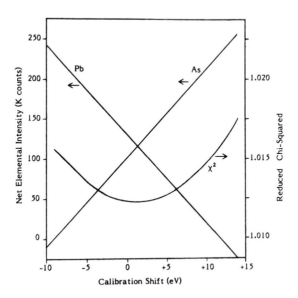

Figure 10.2 Variation in fitted intensity values for Pb and As peaks
as a function of calibration shift (Russ, 1981b).

121

In summary of the errors introduced by spectrum processing, we must cite as potential sources or difficulty: 1) spectra with closely spaced line overlaps, especially where one peak is much larger than the other, 2) possible shifts in relative peak height for one element due to differential matrix absorption, 3) inadequately modelled peak shapes (or changes in the shapes with energy, count rate, electronic interference, etc.), 4) calibration errors of even a few electron volts, and 5) peaks in regions of the spectrum where the continuum is highly curved (e.g. below 3 keV.), or close to absorption edge discontinuities. For major, isolated lines such as K-α peaks from transition metals, where the background is smooth and easily modelled, and overlaps are minimum, an accuracy better than 1% is attainable by most methods. This can easily deteriorate to 10% or more error if the factors above are present.

Correction Models

The Z-A-F model already introduced for converting measured intensities to concentrations has, for most purposes, the best accuracy of any of the models. The others are generally less developed, meaning that there has been less intensive testing to accumulate comparison data, and less exploration of alternate functions and constant parameters to fit the empirical data. Some of the models are indeed fundamentally approximate. Within the equations describing the effects of electron penetration and X-ray emission and absorption in material, there appear several fundamental parameters such as mass absorption coefficients and fluorescence yields. These are not trivial to measure, and good data have been published for only a fraction of the periodic table. Figures 10.3-10.5 summarize the variation in some of the published data.

Estimated absolute accuracy of present mass abs. coefficients

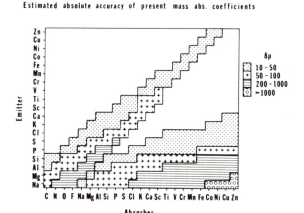

Figure 10.3 Estimated absolute error in mass absorption coefficients (greater errors exist for higher atomic numbers) (Reed, 1971).

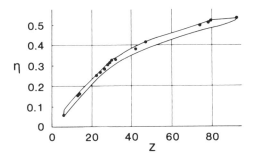

Figure 10.4 Range of published backscatter coefficients

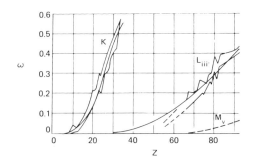

Figure 10.5 Range of published data for fluorescence yields

    The fitting equations used to obtain parameters for missing elements, shells, energies and so on are likewise imperfect. For the K-shell interactions of elements from about Z=13 (aluminum) to 30 (zinc), the very best data exist. Consequently the best accuracy for the computed correction terms may be expected in those cases. The correction terms are themselves only approximate, either using expressions that fail outside a certain range (e.g. $F(X)$), or which omit certain effects altogether (e.g. continuum fluorescence). We can roughly predict when this will produce significant errors in final results. For the case of analysis using K lines from the optimum elements just mentioned, the Z-A-F model is capable of better than 1% accuracy. This can be established by comparing published results from microprobe (WD) analysis in which the other sources of error are small. For L line analysis, accuracy better than 5% is often achievable. The greater error is due to ignoring subshell effects, and limited data for μ and ω (particularly for the rare earths). For M line analysis the situation is even worse, and 10% accuracy is not unusual. The μ and ω value uncertainty, partitioning of excitation among the subshells, and importance of continuum fluorescence for these heavy elements are all important. So also may be the errors in $F(\underline{X})$ for large X values if low energy M lines are strongly absorbed in a heavy matrix. For low energy K lines (the analysis of atomic

123

numbers **4** through about 11 - Be through Na is discussed in a subsequent chapter), the situation is even worse. There is great uncertainty in the mass absorption coefficient values, the $F(\chi)$ term is often quite large, and in addition errors from peak shape distortion, uncertain background shape, and hard-to-define instrument parameters grow in importance. It is unusual to consistently obtain accuracy as good as 10% in this range.

For other models, such as thin section, coating and particle analysis, the inherent limitations of the assumptions made are much greater than for the Z-A-F method. In most of these situations, 10% accuracy may be approached but rarely exceeded. There are important exceptions of course when economic incentives justify the use of special standards and specialized equations for particular measurements, such as film thickness and composition work in the semiconductor industry.

Standards

Standard samples must, of course, be really of known composition, and homogeneous on a fine scale. When pure element standards are used, the error is the same as the purely statistical error in their measurement, which is usually less than 1%. If compound or complex standards are measured and the pure element intensities obtained by a reverse application of Z-A-F factors, the error due to the Z-A-F model must be added as well. The magnitude of these values can be taken from the discussion above.

If ratios of pure element intensity are calculated to estimate one element from another, the error magnitude depends strongly on how far we extrapolate. Within a few atomic numbers, and within one shell (particularly the K shell), the uncertainty in the ratio is very small and the error is not greater than that of the pure element value from which the extrapolation is made. For the L and M shell the results are somewhat poorer, and of course the error increases with the difference in atomic number between the measured point(s) and the calculated one. Nevertheless, within a given shell, the error from the use of ratios will rarely exceed 5% based on agreement shown for measured and calculated data. If an attempt is (foolishly) made to extrapolate across shells, the error is likely to be greater than 10% and may in some cases be <u>much</u> greater. These errors also apply if a "no-standards" calculation is employed with all of the elements normalized together, but because of the normalization the error may be spread around among the elements in ways that obscure its root cause.

Table 10.1 shows an example of the accuracy of extrapolation between nearby elements and across shells (Russ, 1976). The first analysis is of a binary chromium -cobalt alloy. The measured intensities, along with the intensities measured on the pure elements under identical conditions, are listed. These elements are both analyzed by their K lines, but have significant interelement effects of absorption and fluorescence, which require correction. The data may be run through Z-A-F correction in a variety of ways, including the use of both pure element standards, either one, and with none. The values in parentheses show the effect of postnormalization to 100%, which is of course forced in the no-standards case. The error introduced by calculating either element, or even both, is very small.

Table 10.1 Quantitative composition results for Cr-Co alloy using various standards options.

Element	Given Chem.	Meas. Int.*	Pure Int.*
Cr	56.23	4.322	7.176
Co	43.45	1.910	4.927

* counts per second - picoamp, 25 keV.

Standard:	Both Pure	Cr Only	Co Only	None
Cr	57.85 (57.04)	57.40 (57.63)	59.59 (57.66)	(56.98)
Co	43.58 (42.96)	42.20 (42.37)	43.74 (42.34)	(43.02)

() indicate normalized values

A more difficult situation arises when different shells are involved. The second example shows the analysis of a binary gold - copper alloy. Both the gold L and M lines were measured (from the standards as well as the unknowns). This allows even more combinations of choices as to what standards to use. Notice that with the measured pure intensities, quite good results are obtained from Z-A-F correction (the absorption and atomic number corrections are important in this material). The errors due to extrapolation from the copper K to the gold L or M line, or from either of the gold lines to the copper line or to the other gold line, are not small. By coincidence, the normalized results from the no-standards calculation are not too bad; the source and magnitude of the actual errors are hidden by the normalization.

Table 10.2 Quantitative composition results for Au-Cu alloy using various standards options.

Element (Line)	Given Chem.	Measured Inten*	Pure Intensity*
Cu (K-α)	65.0	6.393	9.787
Au (L-α)	35.0	0.891	3.112
Au (M-α)		1.701	6.240

*counts per second-picoamp, 25kV.

Stand:	Both Pure	Cu-K	Au-L	Au-M	None
Cu (K)	64.05 (65.48)	64.06 (67.33)	70.15 (67.40)	85.75 (67.62)	(67.39)
Au (L)	33.78 (34.52)	31.08 (32.67)	33.92 (32.60)	40.08 (32.38)	(32.61)
Cu (K)	64.05 (64.80)	64.07 (70.86)	70.16 (70.15)	85.74 (69.85)	(70.29)
Au (M)	34.80 (35.20)	26.35 (29.14)	29.33 (29.49)	37.00 (30.15)	(29.72)

() indicate mormalized values

In any discussion of the errors introduced by the standards values, it is important to note that if there is any instability of the system, such as a change in beam current that is not measured and compensated for, it appears as an error in the standard pure element intensity value. Variation with time during a lengthy sequence of measurements on standards and unknown, even if they are all placed in the specimen chamber at the same time to avoid irreproducibility in turning the beam off and on again, can easily be greater in magnitude than the potential improvement in accuracy from using the additional standards. The use of compound standards can reduce the number of measurements (and hence the time), as can the use of ratios. This reduction in sensitivity to instrument drift can more than offset the greater error these methods involve. For wavelength dispersive spectrometry, an additional source of error arises from any change in height of the specimen surface, or difference in height between standard and unknown, which changes the X-ray optical focus. For energy dispersive detectors, a change in specimen location or tilt, or average atomic number differences between standard and unknown, can change electron backscattering. These electrons may strike the microscope polepiece or other nearby hardware (the stage, even the sample itself) and produce additional X-ray signals which alter the spectrum.

Conditions

Choosing the wrong analysis conditions or not knowing precisely the ones being used can both introduce errors into the calculated answers. For instance, a 500 electron volt error in measuring the accelerating voltage (or not knowing about a 500 volt retarding field set up by specimen charging) changes the calculated Z-A-F corrections (and also any computed standard intensities) as shown in Table 10.3. The sample is a complex olivine mineral containing oxides of Mg, Si, Ca and Fe, and the actual beam voltage used was 17 keV. The right-hand column compares the variation produced by the same 500 volt error using a correction scheme described in another chapter, which uses two sets of intensity data taken at different voltages (the table shows the results for allowing the value for only one to vary) when the X-ray takeoff angle is unknown. In any case, we see that the combined effect of variation in the assumed accelerating voltage changes the results, due to changes in the ionization, absorption and backscattering terms.

Table 10.3
Sensitivity of calc composition to $\pm$ 500v
error in assumed accelerating voltage

oxide	nominal	conv.ZAF	unk.angle
MgO	48.5	$\pm$ 0.22	$\pm$ 1.32
SiO$_2$	40.80	$\pm$ 0.49	$\pm$ 2.13
CaO	0.33	$\pm$ 0.02	$\pm$ 0.10
FeO	9.80	$\pm$ 0.70	$\pm$ 3.35

A far more serious "voltage error" arises when specimen charging retards the electrons before the strike the specimen. Charging arises when the specimen is a partial insulator and the net specimen gain in electrons (the incoming beam less backscattered and secondary emitted electrons) cannot flow to ground. Charging that deflects the incident electron beam and causes the viewed image to "break up" in the SEM is unstable and must be avoided for microanalysis. But sometimes a high resistance sample will produce a stable retarding field above the sample surface, or even just within the topmost atomic layers. Even retarding fields of many keV. can be present with minimal effect on the SEM image. The lower electron energy will correspondingly alter the production of X-rays. By measuring the voltage from the continuum cut-off in the spectrum, instead of relying on the presumed accelerating voltage from the instrument setting of even from a prior measurement, this source of error can be minimized.

Error related to the electron accelerating voltage also includes the consequences of using too low or high a voltage for the particular specimen. If the voltage is too high, the $F(\chi)$ correction will increase (with attendant errors) and the stopping power calculated from an "average" electron energy will be wrong. If the voltage is too low, the element will be poorly excited (poor statistics). Usually an overvoltage in the range from U=2 to U=7 is recommended, but this is obviously not always possible, depending on the elements present. For instance, even for such a simple combination of elements as Fe and Si, an accelerating voltage of 14 keV. is higher than we would like for the silicon (14/1.84 > 7) and lower than desirable for iron (14/7.11 < 2). Common sense dictates that voltages be selected that adequately excite the elements to be analyzed, but this may increase the absorption correction errors or the lower energy elements in the sample.

Such considerations of overvoltage can also dictate the choice of lines to be used for analysis. Consider for instance the case of Ga-Al-As-P compounds, of considerable interest as electronic materials. Using all K lines for analysis would be preferred, based on previous comments about the quality of fundamental parameters and also because the Ga and As K lines are well resolved and present no peak stripping difficulties. However, if a high enough voltage is used to effectively excite them, the overvoltage is much too great for the Al and P. The $F(\chi)$ value is then large and the absorption correction substantial, and prone to error. Conversely, if we use the L lines of Ga and As, and the K lines of Al and P, good excitation of all elements can be achieved with a reasonable overvoltage ratio. However, the L intensities will be poorly determined because of the peak overlaps between all of the elements. There is no clearcut "best" choice in this instance. With a WD spectrometer, the lower energy lines might be used since they can be adequately resolved. With an ED system, and special software, it might be possible to use different accelerating voltages for the Al and P L lines and for the Ga and As K lines, combining the data into the same Z-A-F calculation. Such degrees of flexibility are not found in most standard programs.

Further complications can arise from the takeoff angle. It may not be possible to position the sample and spectrometer to have a high takeoff angle, for short absorption path length and reduced $F(X)$ corrections, without tilting the specimen surface. This introduces errors in the backscattering correction. Generally, the recommendation that takeoff angle be high for quantitative microanalysis is a good one. But when the angle is of necessity low, the error in its determination can cause significant error in the corrections (principally $F(X)$) that depend upon it. Remember that absorption is exponential with the distance that the X-rays must travel through the specimen, and this distance is proportional to the cosecant of the takeoff angle. The result is a relationship between absorption and takeoff angle that increases dramatically at low angles.

Figure 10.6 shows $F(X)$ values computed using different equations, for aluminum in aluminum oxide, at 10 keV. accelerating voltage. As the takeoff angle drops (the graph covers 4 through 18 degrees, for a horizontal surface, but similar curves are obtained for inclined surfaces) the $F(X)$ value departs farther from 1.0 (larger correction magnitude). Also, the slope increases, and at about ten degrees an error of 2 degrees produces an error greater than 5% in $F(X)$ and consequently in the final composition. This is in fact more important than the discrepancy between the different curves, since by applying the same equation to standard and unknown this will largely cancel out.

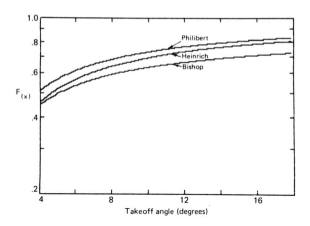

Figure 10.6 Absorption correction for Al in Al_2O_3 at varying takeoff angles, using different equations.

Conclusions

From this description of the sources and possible magnitudes of errors in quantitative electron-excited microanalysis, several lessons can be drawn. Consider, for example:

1) To achieve 1% accuracy (a convenient goal for many users), we must restrict our analysis to major elements whose K lines can be used,

with pure element or high concentration standards (or no standards if the elements all lie close together). A high takeoff angle and well measured accelerating voltage are necessities.

2) The use of standards is not justified unless they can improve total accuracy (when a stable beam or proper compensation for current variations is available).

3) For minor elements (low concentrations) with strong peak overlaps, statistical errors will dominate. The standard intensity for these can therefore usually be calculated relative to the nearby overlapping element with no sacrifice in accuracy.

4) Because of the way the terms add, it is easy for one to dominate the others. It makes no sense to expend money or effort to reduce one source of error unless it is the controlling error. Time spent in accurately measuring analysis conditions may be more valuable than in extending analysis time, for instance.

5) Sometimes the fitting or statistics errors for what would otherwise be the preferred peak for analysis are too great, and another peak should be used instead (either L instead of K, or β instead of α).

Many more specific examples could be cited. If the user will keep in mind the general principles described here, it will usually be possible to straightforwardly assess the relative magnitude of the sources of error and either decide what to work towards improving, or to respect the limitations of the techniques at hand so as to avoid placing undue confidence in the results. Remember that the computer does not know about these limitations and will persist in typing out answers with many digits beyond the decimal point, whether it makes any sense or not.

Chapter 11

Spectrum Processing

When wavelength dispersive spectra are used for quantitative or qualitative analysis, there is practically no spectrum processing to be performed. The narrow peaks cover little background, which can be easily estimated from the level adjacent to the peak or at most by interpolating linearly between the two sides, and overlaps between lines are infrequent. Energy dispersive spectra, on the other hand, require effort to extract the net elemental intensities. The peaks are an order of magnitude or more broader than those from a WD spectrometer, and consequently cover much more of the Bremsstrahlung background. They are very likely to interfere with each other. The "deconvolution" process is often complicated by the presence of extraneous background generated by stray X-rays or electrons, or electrical interferences. The low count rate capability of ED systems often means that the spectra have few counts, certainly for the background and minor peaks, so that counting statistics make the shape of peaks and background "rough"; this can confuse both the eye of the user and the fitting programs. Finally, the detector response is not quite the ideal Gaussian shape that it would be easy to deal with, but has artefacts associated with it that vary from unit to unit and as a function of energy.

Of the various algorithms which have been developed to deal with these problems, some are more generally successful than others, but none is universally preferable. It therefore becomes incumbent upon the user to recognize the particular advantages and limitations of the various approaches, to select the one(s) best suited for the specific job at hand.

Smoothing

Because of the low count rates and statistical fluctuations found in many ED spectra (particularly those from inherently low count rate samples, such as biological thin sections or small particles), many systems incorporate a routine to "smooth" the spectrum. This improves the visual appearance, and so may assist the user in recognizing the presence of trace peaks. It does not, of course, improve the statistical precision of the data, and neither improves the actual detection limits or assists further calculation programs in dealing with the spectrum (in fact, it can with repeated application degrade the spectrum resolution).

The universally accepted method for smoothing of many kinds of spectrum data is to replace each point in the spectrum with a weighted average of itself and its neighbors. The number of neighbors involved and the particular weighting factors depend on the number of points that it takes to cover a peak, or in other words on the spectrum calibration in electron volts per channel since the peaks we deal with will usually be between about 120 and 180 eV wide (depending on spectrometer resolution and peak energy). Savitsky and Golay (1964) have published, and others have added to, sets of coefficients for optimal smoothing. The equation and table below show the method and a few typical examples. The number of points in the smoothing "window" should be roughly the same as the peak width (FWHM), and so will vary from about 5 total (np=2 in the equation) at 20 eV/channel to 9 or 11 (np=4 or 5) at 10 eV/channel, and so on. Note that the factors are applied symmetrically, and that by dividing by the total, the replacement spectrum has the same average height as the original.

$$N_i = \sum_{j=i-np}^{j=i+np} w_{j-i} N_j \ / \ \Sigma \ (w_j)$$

Weighting factors ($w_{-k} = w_k$):

np	$w_0, w_1, w_2,$ etc.
2	17, 12, -3
3	7, 6, 3, -2
4	59, 54, 39, 14, -21
5	89, 84, 69, 44, 9, -3
6	25, 24, 21, 16, 9, 0, -11
7	167, 162, 147, 122, 87, 42, -13, -78
8	43, 42, 39, 34, 27, 18, 7, -6, -21
9	269, 264, 249, 224, 189, 144, 89, 24, -51, -136

If these factors are normalized against their totals and plotted against their fractional distance from the midpoint, they form a universal curve, from which it may be seen that they represent a function akin to an R-C frequency filter. There are other similar filter factors that can be used to obtain derivatives of the spectrum.

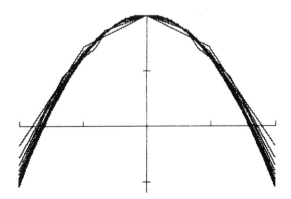

Figure 11.1 Plot of peak filter formed from Savitsky and Golay (1964) weighting factors.

The first method we shall describe for separation of peaks and background is that of fitting of a predetermined functional shape to points in the spectrum, and interpolation under the peaks or subtraction of the function. This is probably the most intuitively clear technique, and in some cases gives quite good results. The simplest fitting function is a straight line. As indicated in the sketch below, if the continuum is curved or has a discontinuity because of an absorption edge (and these will generally lie underneath characteristic peaks from the same element), the straight line interpolation does not do a good job.

Figure 11.2 Linear interpolation under a peak can produce either positive or negative errors in peak area.

However, at moderately high energies (from 4–5 keV. up), most ED spectra have the general background shape of the generated continuum, which varies with energy E as $(E_o-E)/E$ and is little modified by matrix absorption (so the discontinuities are small) or spectrometer efficiency (the entrance window is quite transparent), so that over modest energy spans it can be acceptably fitted even with a simple straight line. Better results are obtainable by using a polynomial function which can follow curvature (but not discontinuities) or an exponential function of energy. The latter is particularly useful at the higher energies because it fits the drop in the continuum fairly well.

A function of the form $(E_o-E)/E$ also works, because it is the fundamental shape of the generated continuum (from Kramers' model). Properly, this should be multiplied by the spectrometer efficiency T and the absorption correction $F(X)$ calculated for the particular X-ray energy for the matrix mass absorption coefficient (based on the total matrix composition), and we will see shortly that this approach is also used. However, at high energies the spectrometer efficiency is constant (usually taken as 1.0) and matrix absorption is either low or at least varies very gradually with energy, and so the simple form of the equation can be applied.

With all these functional forms, it is important to fit over only rather narrow energy ranges, of course. They will not work well if large discontinuities are present, and consequently suffer in accuracy at low energies. And the burden of finding points at which to perform the fit is generally placed on the user. This is probably because these simple models require minimal computing and so are often built into the least powerful (and lowest cost) units, but it is not necessary, since it is possible with a fairly crude algorithm to identify regions in the spectrum that are most likely background, and may be used for fitting. The criterion to be used is that a point is not a background point if 1) it is higher than its neighbor on either side by more than simple counting statistics would account for (it could be a peak), or 2) it is lower than both neighbors (it could be in a valley between partially resolved peaks). In applying this test, or for that matter in carrying out the fitting, we should not use single channels. The counting statistics of the background, in particular, are poor; using regions that cover a number of channels, for instance 100-200 eV., gives much better precision. Proceeding through a spectrum and testing each region against its neighbors, remembering those that are probably background, and then coming back and fitting a simple function to them to interpolate under the remaining regions, requires very little time or computer power. Refinements to limit the minimum span and to ask for user assistance if a preset maximum allowable span is exceeded, add much to the generality of such a routine.

Another technique which is really a 'shape fitting' method lightly disguised, is the fitting of a stored "blank" or background spectrum to the measured one. This may be a spectrum of just continuum measured from a light element target such as diamond, or a previously processed spectrum from which peaks have been removed. It must have been measured at the same accelerating voltage as the unknown spectrum. Since the shape of the generated continuum is presumed to be independent of composition, with only the magnitude varying with atomic number (and the magnitude can be adjusted in the fitting operation), this seems at first to offer a simple way to accurately remove background. However, the measured background shape depends very much upon matrix composition, because of absorption. Consequently this method is also prone to serious error in the low energy range, but often acceptable at higher energies. We can generalize already to expect that practically all methods will work at higher energies, and it is at lower energies in the 1-3 keV. range, where absorption edge discontinuities and background curvature are important, that the serious test will come. Without being unfair to the instrument manufacturers, it is interesting to note that it is usually samples containing transition metals (moderately high energy peaks, not too dense or absorbing a matrix) that they use for demonstration.

Computing the Background

As a direct extension of this last method, it is clear that the background shape can be calculated using the theoretical expression:

$$B(E) = T(E) \ F(X)(E_0-E)/E$$

in terms of the spectrometer efficiency T, the matrix absorption $F(X)$ and the accelerating voltage E_0. With only a single factor to determine the overall magnitude of the curve, fitting could be accomplished from a single point in the spectrum. The result does not fit too well at low energies however, chiefly because the matrix absorption computed with $F(\chi)$ expression that applies to characteristic X-rays underestimates the absorption of the Bremsstrahlung which is generated somewhat deeper in the sample. As we shall see in the next chapter, a somewhat more successful fitting expression is (Fiori,1976b):

$$B(E) = F(\chi)\, T(E)\; (a(E_o-E)/E + b(E_o-E)^2/E)$$

where the quadratic term is purely empirical, but allows the shape to curve down at low energies (where E_o-E is large) to match the observed shape. The two fitting constants a and b are determined by solving simultaneously two equations for two selected fitting regions in the sample. It is usually wise to have one of these at a low energy, say around 3 keV., and the other up in the range where the background is absorbed but little.

If this fitting function is used, it is necessary to recompute the background as a function of composition, as part of the overall Z-A-F iteration process. The widely used FRAME correction program written at the U.S. National Bureau of Standards (Myklebust, 1977) incorporates this method. Of course, it is only necessary to calculate the background under the peaks during the iteration process, but after the final results are obtained the entire spectrum background may be computed for display. An example is shown in the next chapter.

Frequency Filtering

A totally different approach to separating peaks from background relies on the fact that they have fundamentally different shapes, without regard to the physical processes that produce them. The common analogy used to explain frequency filtering is to liken the spectrum to an electrical or acoustic waveform that is being passed through an R-C frequency filter to separate high and low frequencies. The peaks in the spectrum, which rise and fall over a short distance along the energy axis, are the high frequencies in this analogy, and the more gradually varying background is the low frequency. Dealing with the spectrum as a collection of characteristic shapes removes the need to predict where peaks may be located, or where acceptable background points can be found. It also makes it possible to utilize the substantial literature on spectrum processing that deals in roughly similar ways with spectra from other analytical instruments.

The best way to really understand what the filters do is to look at the spectrum in frequency space, for instance by passing it through a Fourier transform operation. Both Fourier and other transforms are used in practice, and the transform can be accomplished by cross-multiplication (convolution) much like the smoothing operation, or by classic FFT (Fast Fourier Transform). The transformed spectrum consists of a series of values that represent the "power" of various frequencies in the spectrum, that is the amount to which they contribute to its overall shape. The sketch in Figure 11.3 shows qualitatively the relationship between the original spectrum (and its separate components) and the transformed result. Since the transformed peaks are principally represented by intermediate frequency information, and the background principally by lower frequencies, they should be separable in this domain. The separation is accomplished by passing a "filter" over the transform; it is multiplied by a function which cuts off or attenuates the undesired frequencies. Then when the spectrum is re-transformed to its original space, only the desired components are left.

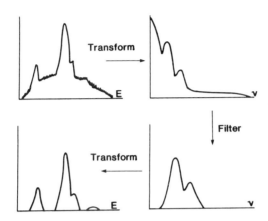

Figure 11.3 Steps in separating the desired frequency components from a spectrum using Fourier or other transforms.

Actually, the smoothing operation described before represents an application of this technique. The statistical channel-to-channel fluctuations represent "noise" that extends up to high frequencies (the contribution of statistics can fluctuate over distances as short as each adjacent channel). Removing it with a filter and re-transforming the spectrum leaves the smoothed result. In fact, the Savitsky and Golay method can be shown to be mathematically identical to the Fourier transform method, and the application of other cross-multiplication or convolution methods can likewise be used to remove other frequency components. When frequency filter methods are used to remove background, they usually remove the statistical fluctuations at the same time, and leave (ideally) just the smoothed background-free peaks.

The problem with this approach is that the frequency domains of the peaks and the background are not all that separate. The peaks have an essentially Gaussian shape which include low as well as high frequency components. The low frequencies in particular are responsible for the skirts on the peaks, and these disappear when a filter is used to cut off all low frequency information, as shown in the drawing in Figure 11.4. Since the integration of peak intensities for quantitative calculations usually does not include these skirts, this would not cause a problem for an isolated peak. But if two peaks are partially overlapped, or worse yet if a small peak is sitting on the skirts of a large one, the clipping of the low frequencies can lead to large errors. This problem can be somewhat alleviated by using the filter to remove the <u>peaks</u> <u>from</u> <u>the</u> <u>spectrum</u> (instead of the other way around), leaving just the background. This can then be subtracted from the original spectrum to obtain the net peaks. However, the background contains some higher frequency information as well, principally associated with the low energy curvature and the discontinuities at absorption edges (smoothed by the detector's resolution, they have much the same frequencies as the peaks). Hence the background obtained by this method is still inadequate for quantitative analysis in the low energy region, and in that case the simpler and faster methods which do no worse might as well be used.

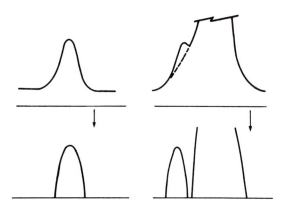

Figure 11.4 Spectrum processing artefacts using frequency filtering.

Before giving up on frequency domain analysis, it is worth noting that the technique of frequency transformation and filtering can also be used to improve spectrum resolution. The detector response function can be at least partially removed to leave sharper peaks, as shown in the example in Figure 11.5 of a spectrum with partially overlapped Cu and Zn K peaks before and after processing. The search for the optimum filter for each spectrum is an iterative procedure, either for sharpening or separation of peaks and background (it depends, for example, on the peak to background ratios), and consequently the amount of computation is too great for this to be a practical technique for routine analysis.

Digital Filters for Differentiation

It was pointed out before that multiplying the spectrum through with a "correlation function", such as the Savitsky and Golay smoothing coefficients, can also be used for frequency transformation or for differentiation. Consider for a moment a very simple function, which is has zero net value and is symmetrical, as shown in Figure 11.6. The values are integer so that quite rapid multiplication and adding is possible, and the processing of an entire spectrum takes very little time. This particular function is used in one of the most common spectrum processing methods, and effectively produces a second derivative of the measured spectrum, as shown in the example.

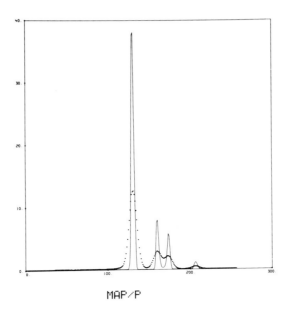

MAP/P

Figure 11.5 Spectrum resolution improvement using an optimum filter to remove detector line broadening (Schwalbe,1981).

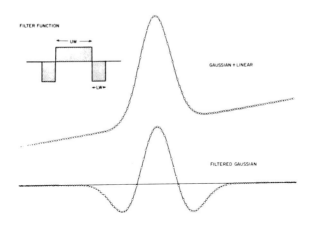

Figure 11.6 "Top hat" digital filter used to remove linear background and find peaks (Schamber, 1977).

The "frequency filtered" or derivative spectrum is not terribly useful by itself, and contains negative values that can hardly be interpreted. However, the differentiation process will eliminate any constant or linear components in the spectrum. This means that to the extent that background could be fit by straight lines, it is gone. This process is applied to spectra before they are fitted with (similarly processed) standard spectra to determine peak sizes. Least squares fitting of spectra is equivalent to saying that the unknown

137

spectrum can be duplicated by adding together, channel by channel, the spectra from each pure element. Only the scaling constants are unknown, and they will be determined in the fitting process. We shall consider shortly how well this works for the peaks. It is clear that it does not work for the background, because the shape of the background depends on several factors. The generation of Bremsstrahlung is proportional to each element's contribution to the mean atomic number of the matrix, which is linear with concentration, while the absorption of the generated radiation is an exponential function in which the concentrations are used to sum up the mean absorption coefficient, and this appears in a nonlinear term. Consequently, we would not expect the continuum to be well modelled by adding together individual elemental backgrounds, and indeed it is not (least squares fitting programs that add complete spectra can sometimes introduce serious errors and are not commonly used for electron excited spectra).

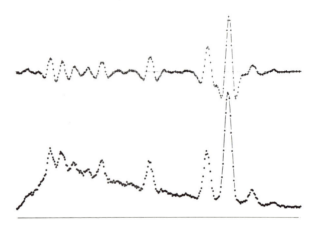

Figure 11.7 An electron-excited spectrum before (bottom) and after (top) digital filtering (McCarthy, 1981).

However, if the background can be substantially reduced in importance in the spectrum, relative to the peaks, then least squares fitting of component spectra to match the unknown may be good enough. That is the purpose of the differentiation process. The "filtered" spectra from pure element standards, or at least portions of spectra that contain only a single element in large concentration, are added together in a linear least-squares process to give a best-fit match to the "filtered" unknown spectrum. It can be proven that if there is no background in the spectra to begin with, the same fitting factors will be obtained using the filtered spectra as the originals. By discriminating against most of the background (the part that can be linearly approximated), enough of the effect of the continuum is removed that adequate results are obtained using this same method.

Peak Deconvolution

In the preceding section of this chapter, the linear least squares fitting approach was introduced. This is at the heart of the most

widely used method for obtaining separate intensity values for
overlapped peaks. The method used is simply to find the "best"
multiplying factors to scale each component spectrum (whether a
differentiated one or one that has otherwise had the background
removed by fitting and subtraction) so that they add together to
closely match the measured or unknown specrum. The addition is
channel by channel, that is for each channel in the unknown spectrum,
we have:

$$N_j = \Sigma \ (a_i \ N_{ij})$$

where j is the channel number, i denotes the various component
spectra, and the a's are the fitting factors. Of course there is no
single set of a values which, when multiplied by the individual in
each channel of each reference specrum, will exactly match the unkown.
The method used to get the "best" values is a linear least squares
technique.

The criterion for best fit is usually to seek the solution (set of
scaling factors) which minimizes the sum of the squares of the
differences between the measured values (channel by channel) and the
summed total of the individual components, times an appropriate
weighting factor for each value (each channel). In most applications
of least squares fitting, for example fitting a linear calibration
curve through curve through a series of points on a graph, the
weighting factor is unity. This is just another way of saying that the
importance or confidence level of each fitting point is the same. But
in dealing with X-ray spectra this is not the case. The measured
spectrum has some channels with very few counts and some with many,
and for each individual channel, normal counting statistics apply.
Consequently, the confidence factor for each number is given by its
variance, and the weighting factors are chosen as N, the number of
points in each channel of the spectrum being fitted. This is the best
estimator available of the "true" or mean number of counts in the
channel, and it can be shown that this criterion produces the most
probable solution. There has been a good deal of discussion over the
appropriateness of using the number of counts in the target or unknown
spectrum as a weighting factor, concentrating in particular on the
effect of very low values where although the total weighting is low
(we place the most importance on fitting where the peaks are, and they
contain larger numbers of counts), the uncertainty in N as a measure
of the true value can bias the weighting. This seems not to make any
practical difference. However, it is clear that when fitting
derivative spectra, the weights should by the intensities in the
original spectrum and not the values of the derivative, which can be
zero or negative.

The fitting criterion has thus become to minimize the quantity:

$$\chi^2 = \Sigma \ ((N_j - \Sigma \ (a_i N_{ij}))^2 \ N_j)$$

where as described the a values represent the fitting factors for each
component, the N values are the number of counts in each channel j of
either the measured unknown spectrum or each of the component spectra,
and the value of χ^2 is chi-squared. The technique is thus also

referred to as "chi-squared minimization" or "weighted least squares fitting". The general class of solution techniques for these problems represents a serious area for extensive study. Fortunately, most of the details can be bypassed in the context of this particular application. The linear nature of the equation shown above allows a solution to be obtained for the a values by conventional solution of simultaneous equations using matrix inversion, easily carried out in a small computer.

The most common application of the method is to fit a portion of the measured spectrum, either in the differentiated form or after discrete subtraction of a fitted background function, to a set of "library" or "reference" specra, one for each of the elements believed to be present. Because of mathematical difficulty and in a practical sense the accuracy of the answer (because the finite number length of the computer will introduce round off errors in the matrix inversion), it is best to keep the number of fitting spectra as low as practical (usually not more than 6 or 8 at a time). Consequently, various regions of a complex spectrum may be fit separately using different lists of reference spectra corresponding to the different elements with peaks in each region. The selection of the lists of elements to be used is generally the responsibility of the user, as the problem of automatic qualitative analysis (determination of what elements are present) is in fact very difficult, and beyond the capabilities of practical computation. (This is beyond the scope of the present discussion, but arises out of the potential for confusion between minor lines of elements with lines of others, the extensive logic needed to decide if all lines of a possible element are really there or if not, if they should be depending on the counting statistics and excitation conditions, and so on. Automatic pattern recognition or search-match programs are used in many fields, including X-ray diffraction, and rarely are able to give unequivocal answers or to outperform the trained eye of a skilled operator in finding a minor component, particularly if the operator has independent knowledge such as the history of the sample, from which to anticipate a likely or possible result.)

The reference spectra are often stored from prior measurements. They may be acquired from pure elements, or from complex samples in which the particular element is high in concentration (for good statistics) and has completely isolated peaks. If the reference spectrum is the pure sample, then the fitting factor will be directly the k-ratio needed as the starting point of quantitative analyses (assuming the conditions, time, etc. of the analysis are identical). The advantage of using measured spectra for the fitting library is that they contain all of the same imperfections as the unknown, in terms of the detector response function. If the particular detector has low energy tails on peaks at low energy, they will appear in both the reference and the unknown spectrum without any need to use elaborate mathematical functions to describe the shapes. Similarly, the library spectrum will also contain all of the element's minor lines, which otherwise would require an extensive table of data.

The problems that arise from the use of this technique are several. First, the storage of many library spectra takes a lot of space, although as larger computers and fast, large capacity disk drives become ever more affordable, this seems less and less significant. Remember that it is not sufficient to simply have one spectrum for each element. The standard spectrum should ideally be measured under the same conditions as the unknown. Changes in voltage or even in surface angle can alter the ratios of peaks in the spectrum, either by changing the excitation of different shells and subshells, or by changing the matrix absorption. Hence, many different libraries obtained under various analyzing conditions must be maintained, and the appropriate selection made. In addition, the line ratios may still be different for the unknown and standard because the matrix absorption depends upon sample composition. The various lines of the element being fitted are at different energies, and the mass absorption coefficient is a function of energy. Consequently, the relative absorption of different energy lines will in general vary from one matrix to another, and in a complex specimen may be quite different from the pure reference standard. This latter problem is particularly important if the matrix of the unknown contains another element with an absorption edge between the peaks of the element being fitted, because then the differential absorption can be quite large. Figure 11.8 illustrates several such situations, many of which involve L lines because of the extended energy range they cover. The probability of this situation arising is not small, since it is primarily when elements have peaks (and hence edges) close together that the need for spectrum fitting to resolve overlaps occurs.

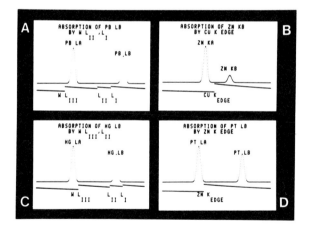

Figure 11.8 Elemental combinations in which absorption edges of one element will cause differential absorption and change the peak ratios in the other element (McCarthy,1981).

Generating Peak Profiles

The other approach to least-square fitting of spectra to match the measured unknown is to synthesize the reference spectra. In practice, most systems that use the fitting method can accommodate any combination of measured and synthesized spectra, since there are compensating advantages and drawbacks. The shape of the measured peak in an energy dispersive spectrum is primarily Gaussian in shape. This describes the major processes in the detector response and the amplification, which were described in an earlier chapter. The Gaussian is fully described by three parameters, the centroid energy, height, and width, as shown in the equation below for the number of counts in a channel j for a peak at energy E, where the spectrum calibration is V electron volts per channel:

$$N_j = H^{-0.5 (E-jV)^2/s}$$

where H is the peak height (at the center) and s is a measure of the width of the peak that is related to the full width at half maximum height (FWHM) by a factor 2.355, and hence can be computed as a function of energy if we know the spectrometer resolution R at manganese K-α :

$$s = (R^2 + 2735 (E-5.894)) / 2.77259 \ 10^6$$

However, as already discussed, the imperfections in the detector add various artefacts to this ideal shape. For fitting purposes, the most important is the low energy tail, a consequence of slow charge collection in surface layers of the device that causes some measured pulses to be too small, and thus introduces counts in the peak that fall below the proper centroid location and cause the asymmetric shape. The magnitude and shape of this tail are a function of energy, and of detector fabrication. Two mathematical descriptions that have been introduced to model the shape of the function are:

1) $N_j = a \ H \ (E-jV) \ e^{-b(E-jV)}$ (Fiori,1976a)

2) $N_j = a \ H \ e^{-b(E-jV)} \ (1-e^{c (E-jV)^2/s})$ (Gunnink,1972)

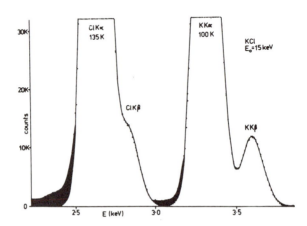

Figure 11.9 Spectrum from KCl showing non-Guassian low energy tails (shaded area) due to incomplete charge collection (Myklebust,1981).

142

They share an exponential shape that describes the overall decline in additional height with distance below the peak centroid. Both are of course only applied for $jV < E$, and represent counts in channel j to be added to the Gaussian shape. The constants a, b and c are totally empirical, and must be estimated for a particular system. They may also be functions of energy. Usually this must be determined beforehand using ideal pure element peaks, and the data or a fitted equation stored for later use. Figure 11.9 illustrates the contribution these tail functions can make in a real spectrum.

The other features of the detector response function must also be considered. The escape peaks can be modelled using the expression for their height relative to each main peak presented before, using a simple Gaussian shape. The broad shelf of counts below the peak is generally low enough relative to the peak height as to be undetectable in an electron excited spectrum, and in any case is likely to be removed from the measured spectrum along with the continuum background. The low energy Compton events are ignored, as are pile up pulses, because they should lie far outside the fitting region that includes the principal peaks. Nevertheless, from the terms that remain it is clear that computing a synthetic spectrum is time consuming, since the process must be repeated for each of the important lines (about 30 lines will cover all of the K, L, M peaks whose height relative to the α -1 line of the respective shell is at least one percent). Usually these energies are available in the system anyway, either in table (Johnson, 1970) or fitted equation form, for use as a qualitative aid, by placing KLM markers on the display. One set of parameters that can be used to estimate rather closely the energies of the important lines (and absorption edges) is listed in Table 11.1. The model is simply a fifth-order polynomial in atomic number. Within the range 0-40 keV. the errors are everywhere small enough for most fitting purposes.

Since the peaks are computed individually in this spectrum synthesis, their relative heights can in principle (this is by no means always done) be computed and adjusted for the particular excitation conditions (e.g. relative heights of lines from different shells and subshells) and matrix absorption. In this latter case, the full matrix composition is required, and so the calculation would have to be placed inside the Z-A-F iteration loop. Considering the magnitude of the computing job needed to synthesize the spectra and perform the least squares fit, this is not practical. Another approach that does not constrain the relative peak heights to a predetermined set of values is to fit each peak individually to the measured spectrum, as though it were a separate component. Then the relative heights can adjust themselves to the actual data, and the reported intensity ratios serve as a check on the reasonableness of what the fit has accomplished. However, this increases substantially the number of components that must be handled, and the number of simultaneous equations and size of the matrix that must be inverted to solve for the fitting factors. In order to handle the mathematics of the process, the fitting regions must consequently be made smaller, so that fewer peaks are included at a time. This may not be possible, if large heavily overlapped regions are present. In any case, it is not at all clear that this offers any advantage over using measured

Table 11.1 Coefficient data table for polynomial fit of line and edge energies as a function of atomic number.

FORMULA: E = SUM (FOR J = 0 TO 5) OF A(J) x Z^J
FOR ELEMENTS WITH Z > MINIMUM VALUE SHOWN

LINE	MIN.Z	A(0)	A(1)	A(2)	A(3)	A(4)	A(5)
K A1	6	.03148688	-.01808771	9.87304E-03	9.765846E-06	-6.512301E-08	2.057676E-09
K A2	20	-.137601	1.235541E-03	9.054276E-03	2.58471E-05	-2.30099E-07	2.194941E-09
K B1	14	-.1399095	-8.785384E-03	.01028758	2.824634E-05	-9.823849E-08	1.358636E-09
K B2	34	9.695049	-1.15356	.06239	-1.134069E-03	1.282856E-05	-5.538106E-08
K B3	31	-1.372427	.1562154	1.630599E-03	2.504917E-04	-2.890185E-06	1.496488E-08
L A1	20	-1.068516	-8.297638E-04	8.848479E-04	1.645521E-05	-1.294784E-07	4.200417E-10
L A2	37	-1.84133	.1456602	-3.932634E-03	9.334239E-05	-7.302407E-07	2.232784E-09
L B1	20	-.2800169	.02122423	-1.938217E-04	4.18068E-05	-3.968281E-07	1.981353E-09
L B2	40	3.839469	-.3239941	.01072464	-1.215595E-04	8.603049E-07	-2.329595E-09
L B3	33	-1.806444	.1611602	-4.957083E-03	1.205541E-04	-1.026203E-06	3.949398E-09
L B4	37	-2.040568	.1744145	-5.154818E-03	1.195479E-04	-9.834283E-07	3.574682E-09
L G1	40	3.120544	-.2552137	8.083908E-03	-7.07888E-05	3.923834E-07	-1.342595E-10
L G2	45	-.07992459	-4.735351E-03	7.211986E-04	3.723517E-05	-3.840286E-07	2.034708E-09
L G3	55	-9.202729	.5812008	-.01411146	2.22012E-04	-1.516948E-06	4.80612E-09
L L	24	-.1603149	.01015415	1.801046E-04	2.728966E-05	-2.105307E-07	4.784643E-10
L N	26	-1.519993	.1390857	-4.613087E-03	1.149816E-04	-9.83775299E-07	3.65824E-09
M A	57	-23.52416	1.52292	-.0392689	5.08724E-04	-3.217463E-06	8.07172E-09
M B	57	-131.2531	9.214725	-.2572545	3.573493E-03	-2.458768E-05	6.722226E-08
M G	41	24.68468	-1.892828	.05683695	-8.29138E-04	6.00739E-06	-1.711003E-08
M Z1	56	-19.89005	1.323146	-.03516256	4.699163E-04	-3.079866E-06	8.0403901E-09
M 2N4	57	157.788	-9.205211	.2140635	-2.472087E-03	1.4261E-05	-3.262014E-08
K ABS	6	.1172787	-.04465722	.01210529	-1.569554E-05	5.861693E-07	-2.04971E-09
L1ABS	20	.4098803	-.04403688	1.657577E-03	3.23073E-05	-4.373702E-07	2.662024E-09
L2ABS	20	.4230414	-.02910859	4.681759E-04	5.696433E-05	-6.651198E-07	3.417335E-09
L3ABS	20	.5017762	-.045319	1.361644E-03	3.436659E-05	-4.109236E-07	1.915817E-09
M1ANS	57	-1.780504	.1455348	-3.909385E-03	5.920714E-05	-4.044693E-07	1.520517E-09
M2ABS	57	1.641887	-.09514456	2.489329E-03	-2.611458E-05	1.757777E-07	-1.539144E-10
M3ABS	57	-5.292596	.3459696	-8.647601E-03	1.119862E-04	-6.567881E-07	1.641959E-09
M4ABS	57	20.27875	-1.759211	.05817274	-9.213963E-04	7.138845E-06	-2.139546E-08
M5ABS	69	-152.2902	7.888253	-.1551049	1.40477E-03	-5.345793E-06	4.859277E-09

144

library spectra with fixed peak ratios, but again performing a series of fits over reduced portions of the measured spectrum which are selected to lie between the major absorption edges of elements present that could seriously alter the peak height ratios.

Calibration Shifts

By far the most troublesome problem with the least—squares fitting method, whether the reference spectra are measured and stored or synthesized from mathematical functions, is the problem of spectrum calibration. If the measured spectrum has a different calibration (electron volts per channel and zero starting point) than the stored spectra, or different from the assumed values used to compute the synthetic peak shapes, the peaks will not quite line up. The fitting operation will still be performed and answers returned, but· they will in general be quite wrong, and there may or may not be any indication that something is wrong. Usually the value of the χ^2 sum for the fit is shown to the user to help him decide if the fit "worked". Large residual χ^2 values can be indications of missing elements from the fit, gross miscalibration, or other such difficulties. However, examples have been published in which small misalignments of the fitted peaks with the measured spectrum have produced acceptable values, and in a few cases even <u>better</u> values of χ^2. The intensities, however, were quite wrong.

This is a troublesome problem. We are willing to accept that the fitting method may not work in all cases, but we would like it to fail in a way that warns us of the failure so that the data will not be accepted and used for further quantitative calculations. A more general statement of the problem is this: Does the "best fit" criterion, by which we demand that the sum of the scaled reference spectra closely match the measured one channel by channel, necessarily lead to the "right" answer in which the scaling factors are related to elemental concentration? The answer is selected instances has been shown to be NO, and this increases our concern for the general case. A thoughtful and moderately exhaustive analysis by Schamber (1981) suggests that the root of the problems observed lies in the statistical nature of the data, and that the propagation of errors is such that when closely overlapped peaks and/or small miscalibrations in peak position are present, the minimization of χ^2 still leads to the proper answer <u>on the average</u>, but the error associated with any individual measurement may increase greatly. The best way to test for this is to repeat the same analysis a few times and process each spectrum. If the results show a variation in accordance with that expected from a simple statistical consideration, then the answer is probably "right". If not, then a sensitive situation has been found which may call for the use of another method, or analysis of the elements by other, non-overlapped lines.

When spectrum calibration may vary from the nominal, or from that used when library spectra were recorded, special additional steps are needed. One relatively simple procedure that is particularly suited to the synthesis of fitting spectra, is to locate two large, isolated and

145

well separated peaks in the spectrum, and from their measured centroid locations (using the computer to assist in finding them), compute the actual spectrum calibration. This may then be used instead of the nominal calibration when generating the peak shapes. It is much less easy to adjust stored spectra for calibration shifts, but by weighted interpolation between channels, a sort of off-center smoothing function, a shifted spectrum can be obtained that matches the calibration of the measured spectrum. It is helpful in this regard that most miscalibrations are zero shifts (displacements of the entire spectrum sideways) rather than gain shifts. The problem arises in the very low signal levels in the electronics and the effect of even very small shifts in ground voltage.

A more general method would be to fit peaks to the measured spectrum in terms of not only their relative height, but also their position. The criterion of goodness of fit is still the minimization of χ^2, but more variables have been added. Since we have perhaps several thousand channels (and certainly a few hundred in the fitting region), handling a few extra variables should be possible without running out of degrees of freedom in the fitting operation. But unfortunately the introduction of the peak centroid energy, or other adjustable parameters such as its width or the a, b, c values that define the shape of the tail function, make the equations nonlinear. It is no longer possible to find the solution analytically by mathematical matrix inversion.

There are, as mentioned before, many techniques for χ^2 minimization. To appreciate what they do, consider the sketch in Figure 11.10. The surface represents the magnitude of the χ^2 term, which is the "error" or difference between the target data (in our case the measured spectrum) and the mathematical description of it based on the values of the parameters being used (the sum of the individual spectral components). The drawing is limited to two variables, which might be the relative height of two peaks or library spectra, or the location and height of a peak. In practice, there are many more variables which we must include, and the χ^2 value depends on them all. It is hard to draw this n-dimensional surface, but the analogy is direct. We seek to find the point on the the χ^2 surface which is a minimum, because the values of the parameters at that point give us our "best" answer.

Since the answer is not known, it is necessary to start somewhere and search for the lowest point, trying to find it efficiently (clearly we can't evaluate all possible combinations of parameters) and avoiding false or local minima which may be present in the surface. Among the standard methods used are: 1) grid-search, in which χ^2 values are computed as one variable is changed until a minimum is found, then another variable (only) is varied stepwise to find a new minimum, then a third, and so on until (hopefully) the true minimum is located, 2) gradient search, in which enough combinations of variables are used to compute χ^2 values and define the slope of the local surface, so that a new estimate can be made downhill, and the process repeated until the minimum is found, and 3) sequential simplex. Since this latter method has actually been used for ED spectra, it deserves a little fuller explanation.

146

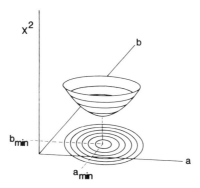

Figure 11.10 Schematic drawing of a X^2 surface for two variables, with a minimum at an unknown location.

Considering the two-dimensional case (for ease of visualization), we shall calculate the X^2 value for three points, reasonably close together and forming a triangle. One of these three will have the worst (largest) value of X^2, so we will look in the opposite direction, by drawing a line through the opposite side of the triangle to locate a new point as shown in the sketch below. The procedure is repeated, with provision for reducing the size of the step if a reversal in direction takes place, so as to be able to move right in on the minimum point. Additional refinements that adjust the step distance in proportion to the degree of improvement in X^2 may also be added.

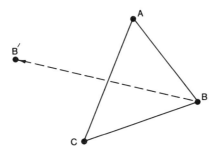

Figure 11.11 A two-dimensional simplex. If point B has the worst value of X^2, then the next point chosen is point B'. The procedure is repeated until an optimum value is reached.

In the general case of more than two variables, the figure is like a triangle, but has one more vertex than there are variables. This figure is called a simplex, and the general use of the method for nonlinear problem solving is well documented. The complexity of the sequential search using a simplex, and the time needed to arrive at the "best" solution, as well as the likelihood of falling into potholes of false or local minima, increase with the number of variables. There is great incentive, therefore, to place as many constraints on the problem as possible. For X-ray spectra, reasonable constraints include upper and lower limits on the width of the peaks (after all, we know at least approximately what the detector resolution is) and on the parameters that define the shape of the low energy tail (they can not have changed too drastically from earlier measurements). Also, while the peak positions may not be known absolutely, they should be fixed relative to each other. Likewise, relative peak heights can be allowed to vary, but within reasonable limits. The computing job can thus be brought within the capabilities of a small computer, and the time required nearly within the range of practical use. Tests on difficult spectra using this method have been very promising (Fiori,1981b), and it is expected that more use will be made of this technique in the future.

Overlap Factors

Another approach to correcting for peak overlaps, which is much faster than the fitting methods, relies on the use of simple factors to describe how much of the intensity recorded at the energy of a line of one element is in fact due to another, and should be subtracted. As shown in the drawing of Figure 11.12, the total number of counts in an energy band ("window" or "region of interest") located with its center approximately on the energy for a particular element (constrained by the fact that peaks occur at energies that may not exactly fall on channels) may include some contribution from peaks of other elements. The overlap may be due to a different peak than that used for measurement of the interfering element, or it may be a portion of the principal peak itself. In either case, if we can calculate the ratio of the fraction of the intensity emitted by the interfering element that lies in the integrated band for the element of interest, to the fraction that is recorded in the principal energy band, then a simple multiplication and subtraction will get us the net intensity of the element of interest.

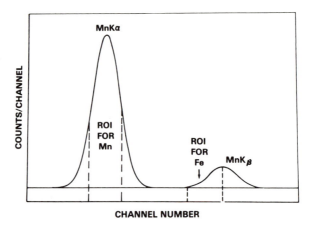

Figure 11.12 Overlap correction is needed for the portion of the Mn K-β peak lying in the integration window or region of interest for Fe K-α. This count can be taken as a fixed or calculable fraction of the counts in the Mn K-α region (Myklebust,1981).

One rather simplistic way to proceed is to set up a fixed series of energy windows for all possible elements, measure a lot of spectra from pure elements (or at least use standards where peaks from any other elements can be accounted for and ignored), and after background subtraction, determine the intensities recorded for all of the absent elements. If these are expressed as a ratio to the intensity for the element present, a series of overlap factors can be obtained as shown in Figure 11.13.

overlapping element	F	Na	Mg	Al	Si	P	S	Cl	K	Ca	Sc	Ti	V	Cr	Mn	Fe	Co	Ni	Cu	Zn	Zr	Ba
F	100	2.10																				
Na	2.01	100	0.70																			
Mg	0.55	1.62	100	0.21																		
Al	0.35	0.46	1.79	100	0.12																	
Si	0.29	0.23	0.30	0.90	100	0.19																
P	0.96	1.12	1.55	2.15	4.13	100	0.70														80.14	
S	0.70	0.67	0.77	0.99	1.40	2.81	100	1.16													3.44	
Cl	0.55	0.47	0.43	0.46	0.59	0.90	1.96	100	0.10												1.11	
K	0.49	0.44	0.32	0.22	0.11	0.14	0.24	0.34	100	7.16												
Ca	0.45	0.35	0.26	0.12	0.08	0.12	0.19	0.33	0.63	100	10.03											
Sc	0.38	0.31	0.20	0.06	0.02	0.03	0.07	0.13	0.23	0.51	100	12.35										83.70
Ti	0.34	0.28	0.17	0.03		0.02	0.05	0.10	0.20	0.40		100	13.70									0.32
V	0.30	0.24	0.14	0.01		0.01	0.04	0.08	0.17	0.32			100	12.65								0.12
Cr	0.25	0.20	0.10	0.01			0.03	0.07	0.15	0.26				100	11.10							
Mn	0.22	0.16	0.07	0.02			0.02	0.06	0.13	0.21					100	8.20						
Fe	1.29	1.50	0.23	0.03			0.02	0.06	0.13	0.18						100	5.00					
Co	1.26	1.10	0.30	0.05			0.02	0.06	0.12	0.15							100	2.50				
Ni	1.24	0.89	0.39	0.05			0.01	0.05	0.09	0.13								100	1.30			
Cu	1.18	24.51	0.61	0.10				0.03	0.06	0.10									100	0.71		
Zn	4.26	282.1	2.07	1.57					0.02	0.07										100		
Zr	0.69	1.14	1.59	2.08	4.69	80.43	5.05	0.13													100	
Ba	1.08	1.92	0.88	0.39	0.25	0.25		0.07	0.33	2.03	95.63	33.38	2.99	2.19							3.71	100

Overlap coefficients expressed as percentages of measured intensities (1.2 FWHM) in analysis peaks of overlapping elements. N.B.: where L line overlap coefficients are expressed as percentages of associated K lines, full differential ZAF corrections are applicable. Coefficients in this table are applicable only at the operating voltage used (15 kV) and, strictly, only at the average counting rate of 3000/sec (full spectrum) at which they were obtained. Changes in counting rate producing significant peak broadening will change the coefficients. Also, they are applicable only to a particular detector resolution.

Figure 11.13 A table of overlap factors (Smith,1976).

These factors are, of course, totally unique to a specific system, detector resolution, window settings, and so on. They also depend on the excitation conditions used, just as the ratios of peak heights in the stored library spectra discussed before did, and will show similar errors if changes in the matrix composition from standards to unknown produce different absorption of the principal and overlapping lines. They do, likewise, include without our specific need to be aware of them any minor lines, escape peaks or peak shape factors.

If we wish to calculate overlap factors, as functions of excitation conditions and matrix absorption, and including all of the lines of possible concern, then once again this must be included inside the overall quantitative iteration loop (since matrix composition is important). This is still in many cases as fast or faster than spectrum fitting with either stored or synthesized spectra to obtain intensities before the Z–A–F calculations begin, and so this has become another fairly widely used method for handling the overlaps.

Special note must be taken of the case in which two elements mutually overlap, as indicated in Figure 11.14, or more complex situations where several elements have mutual overlap factors. It is then necessary to iteratively solve for the net elemental intensities I, such that the equations

$$I_i = T_i - \Sigma\, (OV_{ji}\, I_j)$$

$$I_j = T_j - \Sigma\, (OV_{ij}\, I_i)$$

149

are solved for both (or all) the elements involved in terms of the measured total intensities T and the calculated overlap factors OV.

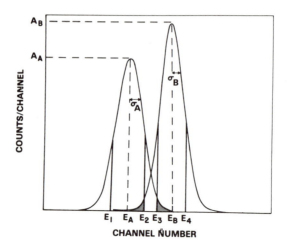

Figure 11.14 Mutual overlap of two peaks, showing the contribution of the tail of each peak into the integration region of its neighbor (Myklebust, 1981).

Also, when overlap factors are calculated, complete equations for the peak shape must be used to obtain the fractional area within the integration regions, and there is the same dependence on knowing the actual spectrum calibration. As before, it is often practical to determine it from the position of two isolated, known peaks in the measured spectrum. The main limitation in accuracy of the method at this time is for overlap ratios that involve different subshells (for L or M lines), or worse yet ratios of lines in different shells, such as subtracting the molybdenum L peaks from an overlap with the K peak of sulfur, based on a ratio to the measured size of the molybdenum K α peak. First of all, it is clear that for this method to be used, we must be able to excite the Mo K shell, which requires the use of higher voltage than would be desireable for the sulfur. But beyond that, the shell -to -shell ratios of intensities are not accurately calculable from fundamentals, and so empirical expressions have been fit to limited sets of measured data, and consequently show 5 to 10% errors in many cases. A 5 % error in the ratio for Mo might result in a much larger error for the net sulfur peak, if it is small. It is hoped that better expressions and more data will improve the accuracy of this rather straightforward and rapid method.

Chapter 12

Background in Electron Excited Spectra

   In the preceding chapter, we saw that practical separation of peaks from background can often be carried out using purely empirical methods that take advantage of the different shapes of peaks and background or fit curves to the background shape found in the spectrum. These are particularly useful when semiquantitative work is being performed, or when the sample is not a classic bulk material for which the semi-theoretical models apply. If these are to be used, they must be incorporated within the overall structure of iterative computation of sample composition because the amount and distribution of continuum (Bremsstahlung) radiation depends upon the matrix composition, as well as upon the voltage and surface orientation. Furthermore, the formulae apply only to bulk materials, and so cannot be properly used with thin sections or particles (although they are sometimes used there, as another example of fitting some kind of curve to the measured spectrum). To understand the model most often used for the removal of background in quantitative microanalysis, we shall start with the understanding available 60 years ago.

   Kramers(1923) derived a relationship for thin targets struck by electrons that stated that the distribution of energy (not number of photons) in any frequency interval was constant for a given material, or that :

$$i_u = \text{const. } Z^2/E_O \, N \, \rho \, dx/A$$

where N (Avagadro's number) times density (ρ) times thickness (dx) divided by atomic number (A) gives the number of atoms encountered on the way through the foil target, and the dependence of cross section is solely on voltage E_O and atomic number Z.

   After the development of quantum mechanics, Kirkpatrick and Wiedmann applied the theory of Sommerfeld to obtain somewhat different results. Their curves, shown below, are not flat (as Kramers' results would predict) but show a variation with energy and somewhat differently shaped curves for different Z.

   If in addition, account is taken of the screening effects of inner shell electrons, which reduce the intensity at low energies (which come from encounters where the electron passes far from the nucleus and is deflected only slightly, and which therefore should involve these inner shell electron clouds), the curves are further modified and become more distinct for different atomic numbers. But this is still a thin foil model, and in strange units. We want number of photons per unit energy interval, and some integration to describe bulk materials.

151

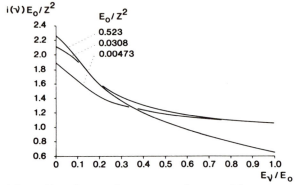

Figure 12.1 Distribution of energy in continuum production by electrons (Reed, 1975b).

It is worth noting that in the thin foil, Bremsstrahlung production is not isotropic (characteristic X-ray emission always is, because the ionized atom loses all history of how it got that way). The moving electron is initially travelling down (for instance) through the foil, and it is very unlikely that the Brem X-ray will be emitted in the backwards direction. At low electron energies (e.g. 10 keV.) the radiation is most probable to occur at 90 degrees, perpendicular to the electron path, which agrees with our intuition about the force needed to deflect the electron trajectory. At somewhat higher energies, typical of the TEM and STEM, the electrons behave relativistically (50 keV. electrons travel at about 40% of the speed of light). This causes the most probable direction of emission to bend forward, as shown in Figure 12.2. Note that the energy distribution is also different in different directions. For the STEM or TEM analyst, the consequence of this is that placing the X-ray detector above the sample improves the peak to background ratio. Of course, there are practical constraints such as mechanical limitations on the space available, shielding from stray electrons and radiation, and so forth, and it is often necessary to mount the detector at the 90 degree position.

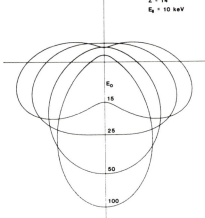

Figure 12.2a Angular distribution of 10 keV. continuum radiation from a thin Si foil at various acclerating voltages.

E$_0$ = 100 keV
Z = 14

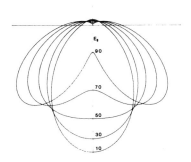

Figure 12.2b Angular distribution of continuum radiation at various energies for a 100 keV. accelerating voltage.

E$_0$ = 15 keV
E$_t$ = 10 keV

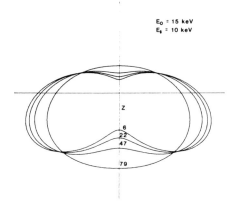

Figure 12.2c Angular distribution of 10 keV. continuum radiation with a 15 keV. electron accelerating voltage, for different atomic number targets.

The drawings in Figure 12.2 are all normalized to have equal areas in the plots. The total continuum intensity from a thin foil is proportional to atomic number squared, although in a bulk sample in which the electrons produce continuum as they drop in energy, the overall proportionality is to atomic number. Note too that the angular distribution is nearly independent of Z (at higher accelerating voltages, the curves of Figure 12.2c become indistinguishable).

For a solid sample, it is necessary to integrate along the electron path using some appropriate model for energy loss. If the Bethe dE/dx expression is applied, we find that the proportionality of intensity is to $Z^{1.3}$ rather that to Z, as predicted by Kramers using a simpler energy loss expression. However, if the reduction in intensity due to backscattering as a function of atomic number is taken into account, the net result is to offset the increase in exponent and have

153

rough proportionality to Z. In terms of the shape if the generated background (number of photons as a function of energy, as we would record it in an ED spectrometer), the simple Kramers derivation gives:

$$B(E) = a\ Z\ (E_O - E)/E\ dE$$

This is, of course, the generated energy distribution and not the emitted (or detected) shape. The equation above rises without bound at low energies, but the low energy photons are absorbed in the sample and the spectrum we actually measure drops at low energy. The absorption depends on the depth distribution of continuum production. If the i_v versus E_v/E curve from above is used as a cross section, then as the electron proceeds along its path the energy E decreases, the ratio E_v/E of the Brem photon energy to electron energy rises, and since the intensity is proportional to $1/E$ the cross section also increases. This means that as the electron is slowed down toward a particular energy, the probability of producing a Bremsstrahlung photon with that energy increases continuously. This is not the same behavior as that for characteristic X-ray production, since the cross section there is proportional to $\log(U)/U$, and below overvoltages of about 2 this decreases sharply. The curves in Figures 12.3-4 show this difference in terms of electron voltage and also as it translates to the position along the electron's trajectory (Rao-Sahib, 1972).

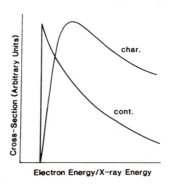

Figure 12.3 Variation of cross section for characteristic and continuum X-ray production with electron energy.

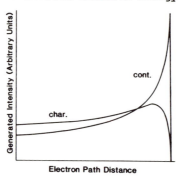

Figure 12.4 Relative generation of characteristic and continuum X-rays along the electron path.

154

The production of continuum X-rays is thus maximum near the very end of the electon trajectories, and this will on the average produce a greater mean depth of generation for background than for characteristic X-rays. Monte-Carlo calculations are in agreement with this predicted behavior, and give depth distributions such as that shown.

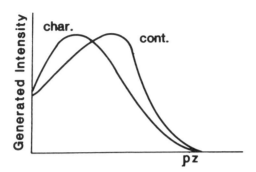

Figure 12.5 Depth distribution - ϕ $(\rho-z)$-curves for continuum X-rays lie deeper in the sample than for characteristic X-rays.

We saw before that the $F(\chi)$ curve used to compute the fraction of generated photons that are emitted from the sample is not too sensitive to the particular shape of the depth distribution curve, and so we might expect to be able to apply any $F(\chi)$ equation to the Kramers shape for generated Bremsstrahlung intensity. This breaks down at low energy, however, because the χ values get quite large and $F(\chi)$ accuracy suffers. Since the background is generated deeper in the sample than the characteristic X-rays, there is more absorption (substantially more at low energy) than predicted by simple $F(\chi)$ curves, and a shape based on:

$$F(\chi) (E_o -E) / E$$

will overestimate the height of low energy background in the spectrum. The low energy X-rays are also absorbed in the entrance window(s) of the detector, further reducing the measured height. Figure 12.6 shows qualitatively the shape of the continuum measured from a bulk sample.

Improving the theory to get better absorption corrections for the low energy end is hampered by the interaction of three factors: 1) the uncertainty in Kramers function itself at low energies, 2) the sensitivity of $F(\chi)$ models to large values of χ , as occur at low energies, and 3) the unknown absolute spectrometer efficiencies at low energies. If the latter are determined by fitting parameters for thicknesses of beryllium, etc. to a continuum spectrum, the values may be nonsense physically (because of the incorrect or inadequate continuum model) but still work fine for practical background fitting and removal because the errors in the thicknesses tend to compensate for the other errors.

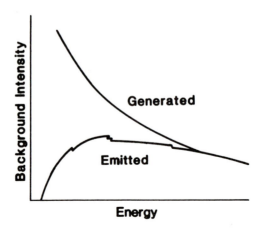

Figure 12.6 Schematic diagram of generated and emitted continuum radiation, showing effect of matrix absorption.

The fundamental expression for background shape, based on Kramers, is just:

$$B(E) = a \, F(X) \, T \, Z \, (E_o - E) \, / \, E$$

where T is spectrometer efficiency, Z is an average atomic number, and a is a constant. For fitting to real spectra, the constant and the Z dependence become unimportant. However, the shape of this simple expression is inadequate to fit measured spectra in the low energy region (below about 3 keV.). Consequently, various fudge factors have been applied:

1) Reed (1975a):

$$B(E) = a \, F(X) \, T \, (\, (E_o - E)/E + T \, F(E) \,)$$

where T.F is a measured and tabulated correction function

2) Lifshin, Fiori and others (Fiori, 1976b):

$$B(E) = F(X) \, T \, (\, a(E_o - E)/E + b(E_o - E)^2/E \,)$$

In this latter expression, the empirical constants are determined by fitting to the spectrum at two background points and solving the two equations simultaneously for a and b. Figure 12.7 shows the fit achieved for a complex spectrum. Note the possibly significant errors at very low energy and remember that the calculation of the absorption correction presumes knowledge of the complete matrix composition.

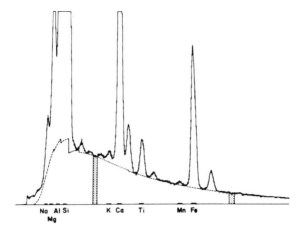

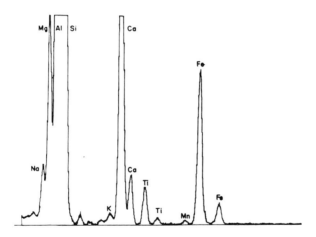

Figure 12.7 Background fit to a complex energy dispersive spectrum as part of a complete iterative quantitative calculation.

Other Considerations

Spectrum background also determines the detection limits that can be achieved. A peak will be detected 95% of the time if it rises above the local background by more than twice the standard deviation of the background, which is just the square root of the mean number of counts. Figure 12.8 shows peaks that are two, five and ten times the two standard deviation criteria. These are very small peaks, and represent sufficiently low concentrations that the elements need not be considered as part of the matrix in fitting the background, so the fitting and removal of the background based on the major elements can be used to detect trace peaks.

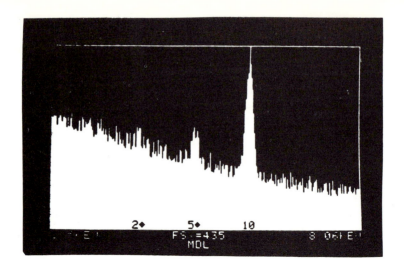

Figure 12.8 Trace peaks rising 2, 5 and ten times the standard deviation of the background count level.

It is also interesting to consider the optimum conditions for maximizing peak to background in bulk samples. This marginally improves detection limits, but is most valuable when making X-ray maps (where the contrast depends directly on the P/B ratio). From relationships developed before, the characteristic intensity is proportional to:

$$C \; \omega \; R/S \; (U-1)^{5/3}$$

while the background is proportional to:

$$Z \; (U-1)$$

so that the P/B ratio should increase with $(U-1)^{2/3}$, which means that very high voltages should give the best peak to background ratios (the same effect is true for thin sections, although the specific efficiency terms are different). The use of high voltages with bulk materials also increases the depth of penetration, which gives more total absorption. But as we have seen, absorption is always greater for the continuum, so this will further increase the P/B ratio. So will the use of low takeoff angles, which again give very high absorption of characteristic and continuum, but preferentially absorb the continuum. The use of low takeoff angles for X-ray mapping is thus another useful technique.

Stray background contributions to the spectrum can be important if they mess up the spectrum shape and impede our ability to fit either the semi-theoretical or even empirical functions. They also increase the background level and make detection limits poorer, and if the stray signals contain characteristic X-rays as well as background, may severely confuse both quantitative and qualitative interpretation. The most common source of stray signals is from backscattered or stray electrons and X-rays in the sample chamber. The drawing in Figure 12.9 summarizes the various signals that can get to the detector: 1) Backscattered electrons that penetrate the window and directly deposit energy in the spectrum, 2) backscattered electrons that strike the polepiece or other hardware of the stage, detector housing, sample stub or holder, or even any part of the sample itself (This usually produces characteristic X-rays as well as continuum, unless low Z materials have been used for shielding. The continuum from this source may completely overwhelm that produced in the specimen for organic or other low atomic number samples, and especially is this true in the STEM where the sample is thin while the support structures are massive), 3) X-rays produced by stray electrons that come down the column outside the central beam (due to scattering from apertures, etc.) that strike other regions of the sample or holder, 4) X-rays produced from the sample, holder and stage by X-rays that are produced in the column (especially at the final aperture, which functions essentially as the anode of an X-ray tube), including X-rays that Compton scatter to the detector (thus appearing at energies that do not match those of the element's characteristic lines).

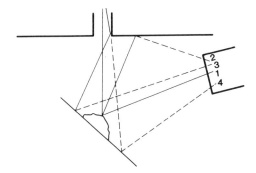

Figure 12.9 Sources of extraneous X-ray signals (see text) (Goldstein, 1978a).

Other sources of spectrum background distortion can include electrical or acoustic noise reaching the detector (try shouting during analysis, or clapping your hands). Sometimes the electromagnetic fields around the microscope lenses and scan coils induce noise currents in the detector (try turning off the scan coils during analysis and look for any differences). Light reaching the detector (a problem with "windowless" detectors since many materials cathodoluminesce under the electron beam) can also produce low energy distortion, in particular. But the most common unrecognized problem is high energy backscattered electrons penetrating through the window, even a normal 7-8 micrometer one, and depositing their energy in the Si(Li) detector. They produce a hump in the spectrum that shows their energy distribution. Indeed, the energy distribution spectra of backscattered electrons has been measured with a windowless Si(Li) detector by subtracting the X-ray signal measured separately when electrons were excluded. For ordinary measurements, the distortion of the spectrum is similar to that shown in Figure 12.10 for an arsenic spectrum. The superimposed line shows the shape of the true

Bremsstrahlung background, determined by fitting at higher energies where there was no electron contribution (because some energy was lost in getting through the window). This background shape would completely frustrate any attempt at fitting semi-theoretical shapes. The situation is potentially much worse in the STEM or TEM because of the higher voltages employed, but fortunately these electrons are often trapped by the field of the polepieces and unable to reach the detector.

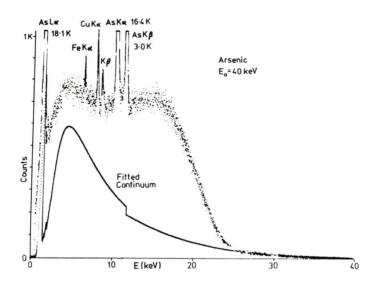

Figure 12.10 Additional spectrum background due to electron penetration to the Si(Li) detector (Fiori, 1981a).

Peak to Background Models

Until now, the spectrum background produced by electron excitation has been a nuisance, and we have devoted no small amount of effort to finding ways to get rid of it. There are some uses which we can make of the background, however. They have to do generally with using it to normalize analyses when we aren't too sure about the excited volume's size or shape. For instance, we shall see that for thin sections the P/B ratio is proportional to concentration regardless of sample density or thickness.

For bulk samples, the only case we have the tools to handle so far is the semi-infinite homogeneous material with a flat surface at a known orientation. Because there are a lot of other common specimen configurations, much work has gone into trying to find ways to deal with some of the more common ones. Two in particular which we shall take up now, are surfaces that are locally rough (with irregularities on a scale about the same size as the excited volume), and particles, which we shall somewhat loosely define as being too large to be treated as thin sections and too small to be approximated as bulk samples (which pretty much means they are not significantly larger than the excited volume). Several methods have been proposed for these cases, but most are highly specialized. One which seems to promise universal applicability is the use of peak to background ratios.

Using intensities from such samples in a normal Z-A-F procedure often produces large errors, exceeding 50% relative to the amount of the elements present. Normalizing the results to 100% not only hides the errors but may actually increase them for some elements. The problem is that the intensities measured from particles or rough surfaces depend not only on the composition, but also on the surface geometry. The absorption path length followed by the X-rays can vary dramatically because of the variations in the surface geometry, and changes in backscattering and fluorescence are also important. For small particles, the through -penetration of the electrons can also affect intensities.

We have just seen that the depth distribution of the continuum X-rays is somewhat different from that for characteristic, that the resulting $F(\chi)$ correction is not identical, and the the stopping power term is different. Nevertheless, it has been observed that to a first approximation, the P/B ratio is relatively insensitive to the size or shape of the specimen, and to the surface roughness. If the stopping power for Bremsstrahlung production were the same as for characteristic X-rays of the same energy, then we would expect this to be the case. The loss of radiation from through penetration or backscattering would reduce the production of both types of X-rays by the same amount, and because the depth distribution would be the same, the absorption of both kinds of X-rays would be the same. Consequently, the ratio of characteristic to continuum (Peak -to - Background or P/B) would not vary, except for the usually minor effect of secondary fluorescence (which cannot generate continuum).

Well, the stopping power may not be all _that_ much different. Or maybe we are just lucky, and the deviations are compensating. On what shall we blame our good fortune? First of all, let us not make the mistake of assuming that this is a quantitative model. The P/B ratio varies less than the intensity, and often a lot less, but still it does vary. Whereas the intensity recorded from different faces of a crystal or particle may vary, for light elements especially, by a factor of 10-20, the peak to background ratio may vary only by a factor of 2. But that still is not in the range of 5-10%. Further corrections are needed.

From earlier presentations we know that the generation of continuum from each increment of path length is proportional to Z^2. We can therefore propose a weighting factor to describe the dependence of background production on concentration for a sample of complex composition:

$$g = \sum_i (C_i Z_i^2 / A)$$

where the A (atomic weight) serves to convert our usual concentration units of weight percent to atomic concentration. The integral of intensity of continuum generated along the electron path can be obtained from Q/S where S is taken as the now familiar Bethe relationship with just J=11.5Z for mean ionization energy, and we use a "universal" curve for Q as an average of those shown before and evaluate the cross section at a mean energy that is the average between the accelerating voltage E_0 and the energy of the generated photon E. Ignoring backscattering, the intensity of generated continuum becomes:

$$I_v = Z / \log(101.4 \, E/Z)$$

Statham (1978) has shown that this agrees rather well with some limited experimental data showing the Z dependence of intensity at a few voltages. Based on this agreement, we shall assume that the stopping powers for continuum and characteristic production are similar enough to cancel in a ratio of their production, except for the "g" factor.

Backscattering is corrected by integrating the energy loss in the differential η versus energy curve for backscattered electrons times the ability of the electrons to generate continuum, proportional to $(E_0/E-1)$, and that for characteristic X-rays, proportional to $(E_0/E_c-1)^{5/3}$. In terms of the bulk backscatter coefficient η this leads to a ratio of the backscatter correction (effective current factor) R for the continuum (subscript w) and characteristic radiation (subscript c) given by:

$$(1-R_w)/(1-R_c) = (2/(1+\chi))^{2/3} \, (0.79+0.44 \, E_w/E)$$

and the total ratio of the "Z" correction between continuum and characteristic is R_c/gR_w.

Similarly, if we increase the magnitude of constant terms in a model like that of Heinrich for the $F(\chi)$ correction, compensation for the added depth of generation should be at least partially obtained. Statham suggests:

$$\frac{F(\chi)_c}{F(\chi)_w} = \frac{1 + 3.34 \times 10^{-6} + 5.59 \times 10^{-13}\chi^2}{1 + 3.00 \times 10^{-6} + 4.50 \times 10^{-13}\chi^2}$$

Since there is no fluorescence of the Bremsstrahlung, no correction is made.

Because of the difference in cross section the P/B ratio is greater for a small particle than for a bulk section, largely due to the increased loss of energy with back, side and through scattered electrons which reduces the background intensity more than the characteristic, and furthermore if the electron scattering within the sample is insufficient to completely randomize the electrons' directions, the anisotropy of Bremsstrahlung production will also reduce the detected background intensity for the normal detector position above the sample.

To actually compute results, Small (1979) describes a sequence in which the first step is to adjust the peak intensity to that which would have been observed from a classic bulk sample:

$$P_{bulk} = P_{particle} * B_{bulk}/B_{particle}$$

and the bulk background value comes from a linear summation in proportion to the concentration of each element in the matrix of the background measured on the pure element (since concentrations are involved, this has to be part of the iteration loop). Then the k-ratio is obtained as peak intensity divided pure intensity for the bulk case. The pure and background intensities can either be measured from standards, or calculated as has been described. The calculation of pure element background is somewhat uncertain because of the inadequate characterization of the atomic number dependence.

Z-A-F calculations from these k-ratios then give concentrations. Published results to date show a marked improvement compared to treating the data as though they came from flat surfaced, bulk samples, but the accuracy does not yet satisfy the need for a universal correction scheme. The table below shows a few representative examples.

Material	Element	"true %"	conv.ZAF	P/B method
Talc	Mg	19.3	10.3	18.5
	Si	29.8	15.8	29.0
FeS_2	S	53.4	39.9	52.9
	Fe	46.6	35.8	46.4
fracture	Au	60.3	28.2	58.0
	Cu	39.6	28.7	44.0

It should hardly be necessary to point out that the use of this method for particles places a burden on the sample preparation, since they must be deposited on a thin substrate which will not generate additional continuum to confuse the interpretation. The further development of these models with universal functions and better accuracy is unlikely to ever approach what is possible with classic bulk samples, but should make possible routine ±10% analysis of particles and rough surfaces.

Chapter 13

Thin Sections in STEM and TEM

Background, discussed in the preceding chapter as a way to normalize intensity variations from irregular surfaces, has another important use (going back much farther in history) in the analysis of thin sections. The background in the spectrum (provided it comes only from the sample and extraneous sources of background have either been eliminated or at least subtracted by making an separate measurement somehow) is proportional to Z^2 in a thin foil. If the average atomic number of the sample is known, or at least unchanging, then the background should be proportional to the total excited mass. This allows measurments on different samples, or different regions of the same sample, to be compared even if the mass thickness (product of thickness and density) vary. Eliminating the stray radiation background is not trivial, although most of the newer STEM's have been designed with this in mind. Earlier methods to record a "blank" spectrum and subtract it from the real one meet with only partial success because the amount of stray background produced by scattered electrons striking hardware near the sample depends strongly on the scattering cross section and mass thickness of the sample, and finding a "blank" sample that contains nothing that will itself generate Bremsstrahlung, but will scatter electrons the same as the real sample is clearly a contradiction.

Peak to Background

From the proportionality of Bremsstrahlung to excited mass comes directly the popular model first proposed by Hall (1974), namely that concentration is proportional to P/B ratio. Figure 13.1 shows the reasoning. The background intensity is linearly proportional to the total mass, the characteristic intensity is linearly proportional to the number of atoms of the element in the excited mass, so the P/B ratio is therefore linearly proportional to the element's concentration. This will be true, of course, only if all of the presumeably linear relationships hold. If absorption of the generated X-rays is negligible, because the dimensions of the sample are such that the absorption path length is too short for significant absorption, and if the cross section for X-ray production (both characteristic and continuum) does not change because the sample is so thin that electrons that have lost energy (which would alter the cross section) have left the sample with no chance for another interaction, then the linearities hold. This has become a working definition for a thin section. Note that a sample may be thin by one criterion and not by both, or that it may be thin for some elements but not others (particularly for very low energy elements whose X-rays may be absorbed in a very short distance, whereas other higher energy elements' intensities are not significantly reduced). Fortunately, the linearity of the P/B ratio often holds approximately for samples that are not truly thin. Indeed, it is because the direction of change in both peak and background intensity with increased thickness is the same for both absorption (increasing thickness increases absorption and reduces intensity) and changing cross section (increasing thickness and decreasing energy increases cross section and hence intensity), that even for bulk samples the P/B ratio is more or less proportional to concentration, as described before.

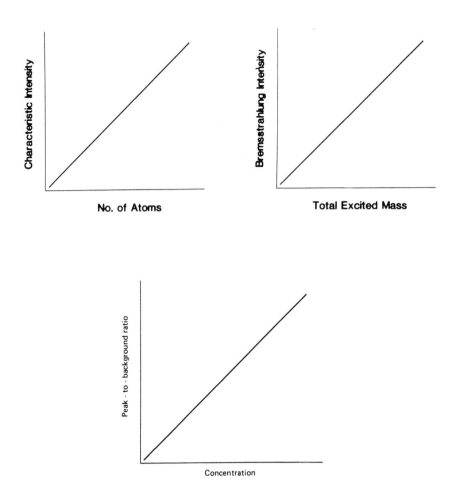

Figure 13.1 Linear relationships in an ideal thin section.

Of course, continuum production is not simply proportional to the mass of the excited volume. There is also a Z dependence that must be taken into account if dissimilar materials are to be compared. This is often done when standards are used that are composed of mineral chips (the known stoichiometry of many compounds makes them useful for this purpose, and homogeneous on a fine scale) or artificial preparations with heavy elements bound in organic molecules. This Z dependence is incorporated in the Hall model shown below (x denotes the unknown and s the standard; the quantity N is the relative number of atoms of the element in the standard, used to get the mean atomic number):

$$C_x = (Z^2/A)_x \frac{(P/B)_x}{(P/B)_s} (N/\sum (NZ^2))_s C_s$$

The terms for the standard can of course be determined. For the unknown, in addition to the measured P/B ratio, we need the average value of Z^2/A. Hall noted that for typical biological materials, this number does not vary too much. For a representative concentration (7%H, 50%C, 25%O, 16%N, 2%S+P) the mean value is 3.28, and this is often used.

The question remains where to take the background. In the previous section, we described peak to background in terms of the ratio of characteristic intensity to the continuum at the same energy. In a thin section, at least two other methods have been used as well. One is to use a region of high energy continuum, away from any elemental peaks. This became common when WD spectrometers were in use, and in order to not tie up the crystal spectrometer and integrate a rather large extent of continuum, a gas proportional counter was used as a crude ED detector to collect just the higher energies. This had the advantage that the although continuum intensity is very low from a thin sample, integration over a broad energy range improved the statistics. Remember that the precision of the P/B ratio depends on the statistics of the background measurement as well. The use of a high energy slice of the continuum makes the implicit assumption that the shape of the continuum is constant, and that only the total amount varies with mass and with atomic number.

Following the same assumption, but using an ED spectrometer, some researchers use the entire continuum intensity. This is generally obtained "free" as a byproduct of subtracting the background from the peaks to get net elemental characteristic intensities. The statistics are obviously as good as they can get, but there is some evidence that the low energy end of the spectrum may vary significantly from sample to sample, particularly when concentrations of heavy elements such as are used in biological stains may be encountered. These represent a serious challenge to the method anyway, both in terms of the criterion of "thinness", which may be true for some regions of the sample but are probably not met in heavily stained areas, and because of the need to know the average Z^2/A value for the region if the equation is to be used. Some programs limit the summation to all background above about 3 keV. to avoid the change in low energy shape. This is probably preferable to using just the background under the peak, because not only does it give better counting statistics, but is less subject to error if the separation of peak from background is less than perfect (and remember that these spectra are statistically poor and may be distorted by extraneous signals). If any error is made in the background subtraction, it will strongly alter the peak-to-local background ratio (anything that increases P reduces B and vice versa).

Fiori (1982) has pointed out that the background in the 4-5 keV. range would be the best choice to minimize the slight dependence in the generated continuum intensity upon atomic number variation beyond that corrected for in the Hall equation (provided of course that there are no peaks or other interferences there). He has also discussed the possibility of computing directly the quantitative relationship between P/B and concentration, but this will require improvements in the models for characteristic and continuum cross sections.

Elemental Ratios

The use of the background for normalization of excited mass is unnecessary when the purpose of the analysis is simply to obtain the concentration ratio of two or more elements in the same region (since the background would cancel out of any ratio). From the same considerations of linearity as presented before, it follows directly that the ratio of intensities of two or more elements should be proportional to their relative weight fractions. The only implied additional assumption is that all of the elements are analyzed in the same volume as each other, which we know not to be true for bulk materials but is true for thin samples. In many cases the measurement of elemental ratios is more meaningful than attempting to determine absolute composition, since the preparation of the sample may have added or removed mass (this may be known for the sample as a whole, by weighing, but not for the local point being analyzed), or loss of mass under the beam may occur. This, and the problem of extraneous background, can be largely ignored using the ratio technique (at least so long as adequate separation of the background from the peaks is still possible).

The simple ratio model simply states that the concentrations "C" and intensities "I" of elements are related by (Cliff,1972):

$$C_1/C_2 = k_{12} \, I_1/I_2$$

where k is a proportionality factor characteristic of the two elements, and the operating conditions (principally the voltage since for a truly thin sample the orientation is unimportant), and the spectrometer response. Since the intensities are measured at the same time and with the same beam current, it is possible to relax the need to measure or correct for current, and even to work with varying or unstable current. The k factor depends on the two elements in the equation and rather than have a table of all possible combinations, it is often more convenient to write the equation as:

$$C_1:C_2:C_3:... = I_1/P_1 : I_2/P_2 : I_3/P_3 : ...$$

where the P factors correspond to each of the elements in the ratio. The P factors describe the relative number of characteristic X-rays emitted from equal concentrations of elements, and thus depend on the detector response and on the electron voltage. They do not, in the absence of non-linear matrix effects such as absorption, depend on the sample. Consequently they may be determined on standards quite different from the prospective unknowns, or as we shall see, calculated from theoretical considerations.

167

Standards

Standards are often used to determine the proportionality factors k or P, and the published curves can even be transferred from instrument to instrument with moderate care in checking the low energy detector response and the approximate agreement of accelerating voltages (at the relatively high voltages used in most STEM analysis, errors in measurement have much less effect because the overvoltage relative to the elements' absorption edge energies are so high). "In-type" standards are materials similar in physical nature to the unknowns. For the analysis of metal foils, the foil as a whole may be able to serve as a standard if the composition of the material is known. Scanning the beam or defocussing it to cover a large area will produce a standard spectrum (provided the thinning process has not preferentially removed one phase) even if the structure is quite heterogeneous, because of the fundamental linear assumptions inherent in the analysis of thin sections. The chief danger in this approach is that absorption may not be negligible in metal foils, and that when large beam areas are used, the higher currents may cause total paralysis of the counting electronics due to the high count rates. But when such a built-in standard is possible, it greatly simplifies the quantitative analysis of local features or inhomogeneities.

Another use of "in-type" standards occurs in the analysis of asbestos-like minerals. These samples are often not thin enough to neglect absorption, but by having a series of k factors for the important elements determined on particles or fibers of different sizes, the absorption and other non-linear effects are "built-in" to the factor and the appropriate one can be chosen for correction of the intensities from any particular measurement. For biological materials, the nearest analogs to the real samples are produced by doping materials such as frozen albumin, methacrylate, epoxies and such with compounds containing Na, K, Cl, and so on. The main criteria for such standards are homogeneity of concentration and stability under the beam. Particularly successful results have been reported for macrocyclic polyether complexes whose structure can trap a variety of elements introduced as appropriate salts. These standards are useful for the P/B or the ratio methods (Spurr,1974 Roomans, 1977).

Ground mineral chips have some merits over organic standards, because they are more stable and have homogeneous, well defined stoichiometry. Figure 13.2 shows a series of sensitivity factors determined at different accelerating voltages from mineral standards. Figure 13.3 shows another sensitivity curve measured with another particularly useful approach to making standards for ratio analysis. This is to spray droplets of solutions containing known (often equal) concentrations of many elements onto a thin film, and measure the intensities to produce a standard curve such as the one shown.

In principle, the P factor for each element can be calculated as the product of terms we already know. It is simply the probability that the electron strikes and ionizes a particular atom in passing through the sample, times the probability that the atom emits an X-ray, times the probability that the X-ray is in the line being measured, times the probability that it is not absorbed (normally this

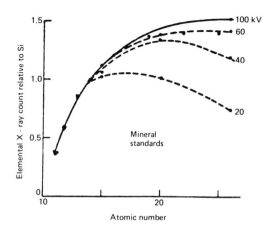

Figure 13.2 Sensitivity factor curves showing relative intensity from equal concentrations of various elements in this standards of minerals at various accelerating voltages.

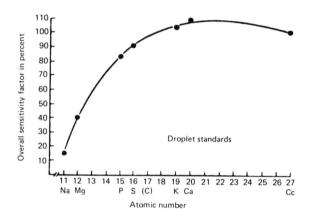

Figure 13.3 Sensitivity factor curves measured from sprayed droplets of solution containing equal amounts of various elements (Morgan, 1975).

is taken as unity for thin sections, but we shall see how to include it when necessary), times the probability that the X-ray is detected and counted. In more familiar terms, it is the product of the cross section Q/A (the atomic weight is as usual needed to get on an equal atomic concentration basis whereas analytical results are by convention given in weight percent) times the fluorescence yield, times the relative line intensity, times the spectrometer efficiency (and perhaps times an absorption term). Consequently we have (Russ, 1974a):

$$P = Q/A \ \omega \ L \ F \ T$$

ignoring any constants of proportionality. Most of these terms we can already calculate, or know where to find (A, ω, L, T). The cross section Q for constant energy electrons is just log (U) / (E_oE_c). Several improvements can be made to this expression. First, at the voltages used in the STEM, relativistic effects are not negligible. The overvoltage should be adjusted to correct for the change in wavelength:

$$U' = U \ (1 + 9.875 \ 10^{-4} \ E_o)$$

which gives about a 10% change at 100 kV. When put into the Q expression, the effect of this change is reduced, and in ratios between elements it is reduced farther, but there is no reason not to include it. Next, we must consider how to include the various shell dependent factors that make it possible to ratio elements measured by different shells. One proposed method (Janossy, 1979) is to add the number of electrons in each shell and some empirical terms that describe the screening effects:

$$Q = A \ N \ \log \ (BU) \ / \ (E_oE_c)$$

where A is 1 for the K shell and 0.714 for the L or M shell (this is effectively the same as using the numbers 2, 5.714, 12.857 for the number of electrons, instead of 2,8,18), and

$$B = 4/(1.65+2.35 \ \exp \ (1-U))$$

Which decreases the cross section at low overvoltages by as much as 10%, which can be important for the analysis of moderately high energy K lines.

Relative line intensities can be determined from tables or measurements from standards, and will not vary since there is no absorption in thin sections. However, because of the problems resulting from overlap of various lines and the poor statistics often encountered because of the low count rates, particularly from biological materials, another approach that is often used is to take the total shell intensity. This is obtained automatically by many fitting programs, and makes it possible to set L equal to 1 in the equation.

The absorption correction, which we would like to ignore, but may be important if the section is "not so thin", can be obtained by integration of the intensity, presumed to be uniform through the section thickness, times the absorption along a path defined by the spectrometer takeoff angle. If

$$m = \mu\rho \ t \ \cos(tilt)/\sin(takeoff)$$

and μ is the mass absorption coefficient for the matrix (the usual linear sum of each individual element's coefficient times its weight fraction), ρ is the density and t the section thickness, then

$$F = (1-e^{-m})/m$$

which is for all practical purposes (when the total absorption is small) identical to the expression (Goldstein, 1976).

$$F = e^{-m/2}$$

that would result if all the X-rays originated at the center of the section instead of being distributed throughout. The remaining problem is to measure the specimen mass thickness (ρt). This can sometimes be done by tilting the sample and measuring the parallax on the top and bottom contamination spots. Another approach is to calibrate the transmitted current or the backscattered signal against (ρt) and the mean atomic number of the sample. When an absorption correction is made, it forces the entire process to become iterative instead of the simple linear calculation we started out with.

Other Considerations

To use these correction procedures, we must first process the spectrum to obtain intensities. Generally the same procedures are used as for bulk material spectra, except that the different shape of the background, and the possibility that it is further distorted by extraneous signals, rules out the semi-theoretical approaches. Instead either filtering out the background or fitting it with polynomials is most often used. For no particular theoretical reason, it often turns out that curves of the form $T (E_o -E)/E$ will fit well over short segments, where T is the spectrometer efficiency estimated with the crude $\exp(1/E^{2.8})$ approximation. If enough background points are chosen to bracket all of the peaks of interest, this or simple polynomials can be used.

One reason to analyze thin sections rather than bulk samples is, of course, to improve the spatial resolution. Instead of excited volumes on the order of a micrometer in diameter, for thin sections the excited region is only a little larger than the incident beam, and can easily be under 1000 angstroms. The transverse scattering of electrons in the sample increases the diameter gradually with section thickness. Attempts to model this with Monte-Carlo require finer step length and calculations based on single scattering events, rather than the more rapid averaging methods that lump together multiple scatters in single approximations. Estimation of possible resolution and measurements across known structures suggest that the lateral diameter in a wide variety of materials will lie between the beam diameter and half the section thickness.

To conveniently estimate the excited volume in thin sections of various materials, it is necessary to account for the variation in electron scattering angles with atomic number. Goldstein (1978b) has proposed that a truncated cone as shown in Figure 13.4 be used to describe the excited volume in a thin section, with dimension b (amount of spreading) given by:

$$b = 0.198 \ Z/E_o \ (\rho/A)^{1/2} \ t^{3/2}$$

in terms of the accelerating voltage E_o in keV., atomic number and weight Z and A, density ρ in g/cm^3, and thickness t and spreading b in nanometers. For an incident beam diameter d this gives a total

171

excited volume of:

$$\text{volume} = t\,\pi\,/4\,(d^2 + bd + b^2/3)$$

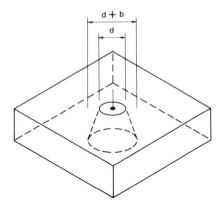

Figure 13.4 Simplified model of excited volume in a thin section.

The dependence of spreading upon the atomic number of the sample and upon accelerating voltage is shown in Figures 13.5 and 13.6. It should be noted that for the greater amounts of spreading, the rather simplified model proposed in this equation breaks down. It assumes only a single elastic scattering event (in the center of the foil thickness) for each incident electron. When the sample thickness or scattering angle become large, multiple scattering events can occur and the amount of increase of spreading will be even larger. In such samples, the basic assumptions about linearity of response upon which our model of thin section behavior is based become doubtful.

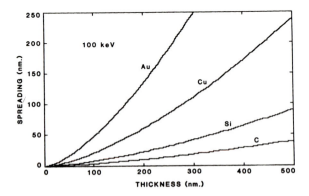

Figure 13.5 Beam spreading with thickness as a function of atomic number.

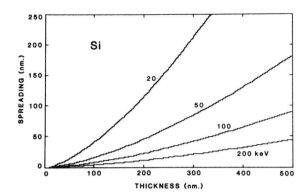

Figure 13.6 Beam spreading with thickness as a function of electron accelerating voltage.

The use of linear equations for peak –to –background ratio as a function of composition are so simple to use, easy to standardize, and appealing to human senses that they have been applied to many other types of samples than ideal thin sections. In low atomic number (eg. organic) matrices, where only a single "heavy" element (in this context, sulfur is a heavy element) is present, good fits for P/B ratio against composition have been reported, and used for routine analysis. And as we have already seen, the rough proportionality of P/B ratio to concentration is exploited for rough surfaces and particles as well.

One particular field of application which is somewhat unique in electron microprobe analysis is the quantitative analysis of residue from evaporated liquids. Nano-and even picoliter volumes of fluids (usually obtained from cells or other organic material) are placed on a very low atomic number substrate, such as beryllium, and carefully dried. The residue is then excited with the electron beam and the X-ray intensities used to quantify the composition using linear calibration curves. The extremely thin residue produces insignificant absorption of the X-rays, and can be adequately treated as a thin sample. Standards are prepared by using fluids of known composition, treated in the same way. Excellent sensitivity and accuracy have been reported for this technique.

Chapter 14

Ultra - Low Energies

This section could really have been titled low-Z elements just as well, since the only real reason for trying to detect these low energies is that the K lines of the light elements B (Z=4) through about Na (Z=11) are at low energies (1 keV. and below). The M and L lines of heavier elements are in this range too, as we shall see, but they are not needed for analysis as other lines are available at higher energy which are easier to analyze. We treat this ultra-low energy range as a distinct topic because it places special constraints on the hardware, excitation, spectrum interpretation and quantitative corrections that must be used.

For wavelength-dispersive spectrometers, there are special crystals with very large 2-d spacings used to diffract these long wavelength X-rays, and the detectors are fitted with extra-thin windows (of great fragility) and often filled with special gases or at special pressures to efficiently absorb the photons and give large pulse amplification. However, the analysis of elements from B to Na with WD spectrometers has been possible for some time, and if it is not exactly routine, at least it requires minimal changes in the hardware.

"Windowless" ED detectors

Conventional Si(Li) detectors are isolated from the vacuum of the electron microscope system by a 7-10 micrometer window of beryllium that absorbs low energy X-rays. It is possible to construct a detector housing that does not use such a window, but instead isolates the detector from the external environment behind some kind of gate valve, which can be completely removed to transmit low energy X-rays to the detector when the sample chamber is at a suitable vacuum. Sometimes the valve itself incorporates a more-or-less conventional beryllium window to permit somewhat normal operation without opening the valve, but these systems are usually bulkier than standard detectors and make it much harder to position the detector close to the sample for efficient X-ray collection, or at a high angle above it to minimize absorption effects.

The truly "windowless" detector configuration is prone to problems. The vacuum systems of most electron microscopes are not clean enough to prevent the deposit of contamination, in the form of both water vapor and oil, on the detector (which you will recall is cooled to liquid nitrogen temperature and acts as a very efficient anti-contamination getter for the sample chamber). Even in microscopes

174

with quite expensive vacuum systems and cooled traps, the contamination introduced by the sample may produce very high partial pressures of these substances in the immediate vicinity of the sample surface and the detector. The lifetime of "windowless" detectors is consequently short, and during that time special efforts are usually needed to remove contamination. (The condensed water vapor can usually be sublimated without harm by warming the detector up in a really clean vacuum environment. The oil molecules are often disassociated or 'cracked' on the surface and much harder if not impossible to remove).

For this reason, most systems are fitted, either permanently or with the capability to be removed for replacement or for occasional 'bare' analysis, with some kind of ultra-thin window that will transmit a reasonable fraction of low energy X-rays, but reduce the transmission of contamination molecules. These films are usually organics, such as parylene (C_8H_8), and since they will not withstand a pressure difference of one atmosphere, they must still be protected behind an isolation valve when removed from the microscope or when the chamber is vented to introduce a sample. The ultra-thin window (UTW) is generally coated with a thin evaporated film of a conductive metal, such as Al (1000-2000 angstroms thick) to block the transmission of light, to which the ED detector is sensitive (this is an additional problem with the completely windowless detector) and further reduce the permeability of the window to contamination. The low energy transmission efficiency is still quite satisfactory, as shown in Figure 14.1. The curves show the transmission efficiency for detectors with a conventional 8 micrometer beryllium window, a 2000A Al plus UTW window, and no window at all (in all cases, the presence of 100 angstroms of gold and 0.2 micrometers of "dead" silicon has been included). The conventional window cuts off detection below about 1 keV (Na K X-rays), while the other two give adequate sensitivity down to boron (other practical considerations limit the ability of these detectors on electron-excited systems to elements from carbon up).

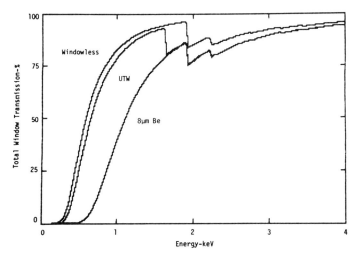

Figure 14.1 Window transmission for conventional and ultra-thin windows on Si(Li) detectors.

175

When such a detector is used, it becomes important to be able to adjust the system parameters in ways that are different from conventional analysis. The amplifier time constant, pulse pileup rejector setting, ADC conversion gain, and especially the detector bias voltage all have important effects on the usefulness of the results. For instance, the thickness of the silicon "dead" layer (the region near the front surface which does not produce electron-hole pairs that are collected to form a pulse, and which therefore can be treated as just another entrance window for X-rays to penetrate to the active region of the detector) depends strongly on the bias voltage, and so different detection efficiencies may result if the bias is not optimized. Contamination of the detector surface also reduces the detection efficiency dramatically. Even the absorption edges from oxygen in water vapor or carbon in organic molecules can sometimes be seen.

The bias voltage also strongly affects the linearity of the output pulse from the detector as a function of photon energy. The total system linearity, or in other words whether the low energy peaks show up where they belong in the spectrum, depends also on the linearity of the amplifier and the analog to digital converter, in a portion of their operating range that is not normally important (and may even be excluded from their specification values by the manufacturer). All of these factors combine to displace the peaks from their rightful positions. Individual systems with less than 3% nonlinearity down to carbon have been reported, but this is hardly typical. Often the use of KLM markers becomes impractical in this range, and the user must resort to "calibrating" peak position by measuring known materials, and even then minor changes in these system components (e.g. due to aging of electronic devices) can alter the nonlinearity unpredictably.

Low energy X-rays penetrate a very short distance into the detector, and so the electrons that are collected can be strongly affected by the behavior of the surface layers. Nonuniformity of bias voltage from place to place (depending on the uniformity of the gold contact layer, also doubtful), trapping or slow collection of these electrons can all produce charge straggling. This has been previously described as the cause of low energy tails on otherwise Gaussian peaks. At higher energies, the tails may account for a few percent of the peak area. At ultralow energies the percentage is much higher. The actual peak shapes recorded depend strongly on the individual system characteristics. They are strongly affected by bias voltage, microphonic or electronically induced noise, by changes in the detector leakage current (the current which flows in the abscence of any incident photons) due to surface or edge contamination build up on the detector, and by light or electrons that reach the detector. This means that even for a given system, the peak shapes are not constant and consequently storing or parametricizing the shapes is at best an occasionally useful method.

Counting rate limitations at these low energies are severe. The pulse pile-up rejectors useful at higher energies function by having a separate amplifier channel with a short time constant process pulses which, although they have poor proportionality of height to energy, can be detected above a discriminator voltage setting to sense the arrival of a pulse and start timing tests for pileup. At very low energies, the pulses are so small that they can no longer be discriminated from normal background electrical noise in such a scheme. Consequently we can forget about pulse pileup rejection. Instead, we must operate at very low total system count rates (remember that it is the total rate of all pulses being processed, including those at higher energy, that determines the probability of pileup). Table 14.1 shows the magnitude of the pileup problem. In addition to the sum peak, there is of course also a shelf of partial sum events that would add to the spectral distortion. Even at extremely low rate of a few hundred per second, there is significant pileup. In no case can the systems be operated usefully at 1000 cps or above (Russ,1981a).

Table 14.1 Ratio - Sum Peak to Main Peak

Elem/Line	200 cps	600 cps	1500 cps
C -K	1.4%	1.9%	3.4%
O -K	1.0%	1.4%	2.6%
Cu-L	0.6%	1.0%	1.7%

Most systems incorporate in the amplifier or ADC a simple discriminator to eliminate or prevent storage of the very low energy counts that result from electronic noise, to keep the rather large peak that would be stored at zero energy from cluttering up the display, and to prevent the ADC's having to spend time processing these pulses. In order to store the low energy pulses of interest for light elements, it is necessary to turn this down so that some storage of the noise peak does occur (in fact , the tail on the noise peak extends up to energies of several hundred electron volts, and this forms an important part of the background in this region of the spectrum). The size and shape of the noise peak can change with microphonic or electrical noise, or if light reaches the detector. The setting of the low energy discriminator may itself introduce nonlinearities or false peaks in the stored spectrum, or by cutting off part of the background, create the appearance of an additional peak to confuse interpretation.

Finally, in considering the hardware of the system, it is necessary to point out again that ED detectors are responsive to electrons as well as X-rays. Problems were noted before when high energy electrons penetrated conventional beryllium windows. With an ultrathin window or windowless system, backscattered electrons will produce far more counts than X-rays (since there are more of them), creating a spectrum in which the dominant feature is the energy distribution of the backscattered electrons and the X-ray peaks are obscured. To prevent this an electron trap, such as a magnetic field from a small permanent magnet, must be arranged to deflect electrons coming from the sample so that they neither reach the detector themselves, nor anyplace that they can produce X-rays that can reach the detector. In fact, the entire detector housing and collimation must be designed to prevent the detector from seeing anything except the sample. The use of knife edge annular collimators and composite materials to absorb each others' X-rays adds further to the bulk and complexity (consequently to the cost) of these devices. The side effect of special electron traps and collimators is that the specimen to detector geometry usually must be fixed, so that various stage heights cannot be accommodated.

Excitation of Low Z Elements

Electron beams in the SEM are not a good tool to excite X-radiation from the light elements. The depth of penetration is rather deep compared to the short distance these X-rays can travel before being absorbed in the matrix, and consequently only those generated very near the surface escape. Figure 14.2 shows the case for oxygen in SiO_2, with a 5 keV. electron beam. The shape of the $\phi(\rho z)$ curve is familiar, but the absorption of the X-rays cuts off the actual emission profile abruptly (even with a 40 degree takeoff angle). For lighter elements and more absorbing matrices, the effect is obviously greater still.

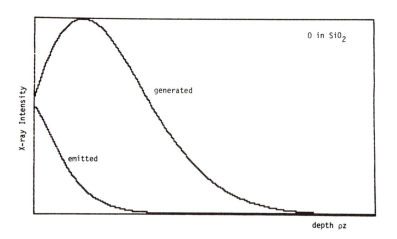

Figure 14.2 Depth distribution of generated and emitted X-rays for O in SiO_2.

Surface excitation by alpha particles or protons is much more efficient than electron because of the shallower range and also the greater cross section, and will be discussed in a following chapter. It is also important to remember that the fluorescence yield of these low atomic number elements is very low, with Auger electron production many orders of magnitude more likely. Much light element surface analysis is therefore done using scanning Auger systems.

Finally, the shallow depth of emission and the easy absorption of these low energy X-rays mean that the presence of surface contamination on the specimen, either before introduction to the SEM (for instance a thin oxidation coating on a metal part), or during analysis (the classic buildup of carbon contamination) will drastically alter the measurements, either by adding low energy peaks to the spectrum from the contamination layer or absorbing those from the sample, or both.

Data Interpretation

In spite of all the problems recited above, it is still possible to record spectra showing peaks from light elements. An example is shown in Figure 14.3, in which the C and O peaks from calcium carbonate are evident, well formed and resolved from each other and from low energy noise produced in the detector and electronics. The question remains: Of what use are these data and how can they be interpreted?

CACO3 (15KEV)   LK Z=20 CA

PR= 119S   119SEC   0 INT

V=8192 H=10KEV 3:4Q   AQ=10KEV 3Q

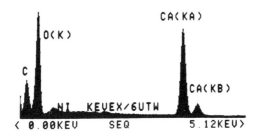

Figure 14.3 X-ray spectrum obtained with an ultra-thin window Si(Li) detector.

Qualitative identification is usually but not always possible. The uncertainty of peak position (due to linearity problems) combined with potential overlaps by the low energy L and M lines may obscure or confuse the spectrum so that positive identification is unreliable. There are very many L and M lines in the region of the C, N and O peaks, and since they are subject to the same problems of distorted shape and shifted position, they often completely hide the peaks from the light elements of interest. Good examples are the L line of Ti hiding the K of nitrogen (thus making identification of titanium nitrides impossible), and similar overlaps of V and Cr with oxygen. Even the escape peaks of higher energy lines can interfere (for instance the sulfur escape peak with oxygen) since the escape peaks are maximum in size at low energy.

Because of these interferences and the variability and complexity of the low energy background, detection limits are poor. Except in unusual cases they will be of the order of several percent and up for electron excitation. The background in this region includes several electronic sources of noise in addition to the Bremsstrahlung, and

179

even the latter cannot be well fit with equations based on the Kramers model because the large absorption corrections and importance of the initial stopping power, when the electon energy is far from the mean, introduce large errors into the approximations used. It is usually only by comparison of spectra from similar samples that reliable estimates of the presence of light elements can be obtained using an ED spectrometer (again, the higher resolution of the WD system improves its detection limits significantly in this region).

Quantitative analysis requires precise net elemental intensities, which except in unusual cases cannot be reliably obtained from an ED spectrum at low energies for the reasons already given. However, if intensities can be obtained (or are taken with a WD spectrometer), they still cannot simply be plugged into the kind of Z-A-F model useful at higher energies. The approximations for stopping power and absorption, in particular, are inappropriate. There are somewhat more elaborate Z-A-F programs that can be used, but it appears that the modelling of $\phi(\rho z)$ curves and application of numerical methods to integrate the absorption may be the most suitable for use with the very low energy elements. The mass absorption coefficients themselves cannot be evaluated from the simple models, but are usually taken from tables.

Finally, it is worth noting that for analysis of minerals or other oxides, experience with conventional WD microprobes suggests that calculating the concentration of oxygen either by difference, or from user-entered stoichiometry, usually gives as good or better results than does the use of measured oxygen intensities with calculated corrections. It is certain that as the intensities from ED systems are beset with additional problems that adversely affect precision, the same conclusion could be drawn even more forcefully in that case.

Several published papers, including this author's, have demonstrated quantitative light element results (principally for C and O, being elements of some economic importance). It can be done, but the constraints on specimen preparation and contamination, counting procedures, spectrum processing, selection of standards, and the use of an appropriate model and fundamental parameters data, make it more of a curiosity than a useable routine analytical procedure.

Chapter 15

X-ray Distribution Maps

Much of the discussion in the preceding chapters has dealt with the possibilities, limitations and methods for quantitative analysis - the determination of the complete composition at a particular location. Frequently the information that is desired is rather what the distribution of an element or elements is in a heterogeneous sample. The ways to answer these kinds of questions have not been developed in as much detail, and certainly with less mathematical modelling, than the quantitative questions.

The technique most used is virtually unchanged from the early development of the microprobe, and is generally known as "X-ray dot mapping". With a WD spectrometer, the crystal is set to the proper angle for a particular element of interest, and then for each X-ray count, a pulse is sent not only to the counter, but also to the amplifier that controls the Z (brightness) of the display cathode ray tube (CRT). Since the X-and Y-position of the beam in the CRT corresponds to the position of the electron beam on the specimen (this is after all the basic principle of the SEM, and the magnification of the images is simply the ratio of scan distance on the display scope to the distance on the specimen), when an X-ray is detected, the pulse produces a dot on the screen in a position that identifies the location on the sample from which it came. Recorded photographically, the resulting image can be compared to electron or other images to study the correspondence between areas rich (or poor) in the element and visible structural features.

Of course, in the electron probe, the possible count rates are rather high, the sample surface is generally flat, and the good resolution produces a very high peak to background ratio. Compared to the performance of WD spectrometers, the problems of E D spectrometers in these particular areas will be discussed below. The ED system relies on electronic selection of pulses in a particular energy range to form an element-specific image. This can be accomplished with a single channel analyzer (a pair of voltage comparators that pass pulses in a particular voltage range), but the most common way to select a particular element with most systems is to select a range of channel addresses in the MCA, called a "window" or "region of interest". Then whenever a count is stored at those memory locations, the instrument generates a pulse for the imaging amplifier. The amplification and digitization process take a little time, of course (perhaps 50 microseconds), during which the beam has moved a little bit, but the shift in the X-ray image compared to the normal electron image is negligible at the relatively low scan rates used for X-ray mapping. Integration of many dots on film, during either a slow single

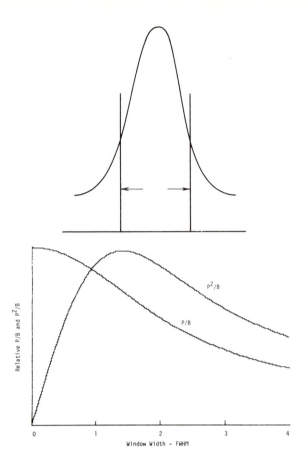

Figure 15.1 Variation in P/B ratio and product of P/B times count rate with width of "window" or "region of interest" used for selection of element.

scan or repetitive scans of the surface, builds up the familiar dot map.

In order to see variations in such a map, which we shall for the moment assume are related to changes in concentration of the element selected, and indeed the ability to distinguish the element's presence from the background level of dots that result from continuum X-rays at the same energy as the characteristic peak, which of course are also stored in the same energy range, we must consider the number of dots on the image and the peak to background ratio of the peak in the spectrum. Contrast between areas where the element is (dots from peak and background) and where it is not (dots from the background only) must be maximized for us to be able to see low concentrations. For large peaks (major elements) this is not too much of a problem even with the ED spectrometer, whose peak to background ratios are about an order of magnitude poorer than the WD type. For minor elements, it becomes very important to consider how the region of interest is set up with respect to the peak, because the peak to background ratio that controls the image contrast is not the ratio of the height of the top of the peak to the background, but the total of all the characteristic counts in the region to all of the background. As shown in Figure 15.1 for a Gaussian peak on a flat background, the total peak to background ratio declines as the window width is increased (the example is for a peak whose centroid height is five times the background level). This would suggest that we should use very narrow windows for X-ray mapping, perhaps just a single channel at the peak's centroid. But the second requirement for the ability to distinguish between the dot density at two different areas on the photograph is number of counts. Simple counting statistics tells us that the more counts we have in a small area, the less is the relative standard deviation and the smaller the real difference there can be for us to find it. So we need lots of counts (dots), and using a narrow window defeats that effort because we are discarding most of the characteristic X-rays that are detected. The better criterion is thus to maximize not simply the peak to background ratio, but the product of the P/B ratio and the total counts. As seen in Figure 15.1, this curve reaches a maximum when the window width is slightly greater than the peak's width at half maximum height.

Some effect on the image contrast can also be produced by choice of the analysis conditions. It is clear that we must use an accelerating voltage, for instance, that will strongly excite the element and thus produce a substantial peak. If the overvoltage is too low, poor P/B's will result. Raising the voltage produces several effects. First, the element is more and more strongly excited (excitation rises, as we have seen, proportional to (U-1) raised to the 1.6 power). The peak to background ratio continues to improve, and this is abetted still further because the continuum background, as we saw, is generated somewhat deeper in the sample than the characteristic X-rays, and is thus more strongly absorbed. Choice of accelerating voltages much too high, from the standpoint of making an accurate absorption correction, or low takeoff angles, can thus be used in some cases to improve P/B ratio and hence contrast in X-ray maps.

However, the total count rate produced from background and other elements in the sample will also increase with accelerating voltage, and this may swamp the spectrometer's counting ability and degrade or even alter results, as we shall see. Also, higher accelerating voltages may produce more scattered electrons or stray X-rays, which are also undesireable. Finally, as the accelerating voltage is raised, the size of the excited volume increases. If the size of features being mapped is large compared to the dimensions of the excited region, then this is acceptable. For studies of grain boundary or diffusion concentration profiles, however, or high magnification images where the size of the excited volume is the limit to image sharpness, it would not be permissible to raise the voltage too high. Generally, then, we will expect to use somewhat higher accelerating voltages than would be selected for quantitative analysis, and can of course change this parameter for different elements as needed, but with the points mentioned above in mind.

The integrated P/B ratios from ED spectra are not too large, and consequently X-ray mapping to show trace elements or small variations is very difficult. Also, the count rate limitations of the systems restrict the number of total X-rays that can be processed, of which the ones actually imaged may represent only a tiny fraction. Hence it often takes quite long scanning times to record a good quality X-ray dot map. Other things being equal (e.g. the same actual elemental distribution), the quality of the map will improve with the square root of the number of counts or the square root of time. Since the quality can be roughly equated to the number of dots (4000 is usually enough for a rough guess with a major element; 20,000 dots give a decent map for an internal report or lab note book scan on a minor element; 100,000 to 1,000,000 dots may be used for a publication quality micrograph), one convenience feature found in many systems is the ability to either monitor the total number of dots counted in the window, or to preset the system to stop at a certain number. In the latter case it is important to either make sure beforehand that the time is long enough for the beam to have completed at least ten or more full rasters across the specimen surface, or to link together the scan circuits with the ED system so that the analysis and dot-generating process will start and stop between consecutive scans (or else a line where the number of dots suddenly decreases will be visible on the photograph). With at least ten scans, the termination line is usually invisible.

In fact, the human eye is not a very good estimator of dot density. It takes really major changes, particularly at low dot densities, for us to perceive the change. Our eye/mind combination is much better at finding edges in images where brightness changes abruptly. One way to convert the dot map to a brightness or grey scale is simply to turn down the CRT brightness so that individual dots leave a partially exposed dot on the film and then accumulate a very large number of dots. Multiple dots in the same place on subsequent scans will cause more exposure on the film, and the individual dots will tend to fade from view, until a grey scale image is perceived. This is not too different from the way a half-tone image used in most printing of photographs in magazines works; you see an overall brightness level and not the individual dots. Other methods for converting dot images to grey scale include some of the computer and ratemeter methods we shall discuss below, and a very crude yet surprisingly effective darkroom technique of rephotographing the dot image out of focus so that every dot becomes a sort of grey halo, and the overlapping of many dots produces a grey scale. Image resolution suffers in this process, but when the only record you have is a dot map with too few dots, you might try it.

The X-ray map or image resolution is not all that great to begin with. The spatial resolution is controlled by the size of the excited volume, which is of the order of one micrometer in a bulk sample, whereas the secondary electron image is formed by the signal coming primarily from a region not much larger than the beam diameter, 100 times smaller. Consequently, for a bulk sample the maximum real magnification possible with an X-ray image is a few thousand times.

It takes a long time to accumulate enough dots for a really first class grey scale image. The natural temptation is to increase the beam current and with it the count rate. By monitoring the total count or dead time of the system, it should be possible to stay withir limits of good system performance, say 30-40% maximum dead time this is often an invitation to disaster. Remember that the s_ count rate or dead time is an average, obtained over the length oɪ time that it takes for the beam to scan over and average some area of the sample surface. But X-ray maps are made almost by definition on samples that vary in composition from place to place. That means the count rate varies, and with it the dead time. In samples with large changes in total X-ray signal from place to place (organic samples with localized concentrations of heavy metals are classic examples), it is possible to find that an average count rate of one or two thousand counts per second while scanning over a large area corresponds to a local count rate when the beam is in a high count region of several tens of thousands of counts per second. The system response is of course to reject those piled-up counts, so the high count rate does not appear on the map. If the element being monitored was not the one that generated the high total system count rate, or even if it was and the count rate went high enough to paralyze the system entirely, the result in the map will be not an increase in counts or even a constant level, but a drop. Figure 15.2 shows an example produced by scanning over a piece of graphite containing two metal wires, one of copper and one of zinc. The first image shows a normal dot map for copper (Figure 15.3 shows the secondary electron image). The second image shows the map obtained at a higher beam current (the "average" count rate as the beamed scanned the entire area was 5000 cps). The dark areas where both wires are located show the paralysis that occurs at high count rates. The third image shows another common problem with this type of image: the window was not set on either peak, but on a region of background. The user might for instance have set it at the energy of another element, such as titanium. The image appears to show a small but discernible concentration of this fictitious element, when actually the increased dot density is simply the consequence of more continuum generation (remember that continuum intensity is proportional to atomic number), and hence more dots produced by storage of X-ray counts in that region of the spectrum.

These effects are obvious when produced one at a time in this way. They may not be so obvious when mixed together in a real sample, and can obscure the real variations in elemental intensity that are of interest. Other problems can occur when a minor peak from some unexpected second element contributes counts in the region of interest. Even an escape or sum peak can create enough additional counts to be mistaken for a trace level of the element of interest. It is important to keep in mind that as the X-rays are counted, there is no way to distinguish a characteristic X-ray of one element from a continuum X-ray or a minor line of another element; they all have the same energy. It is only after a statistically significant total number of counts have been recorded in the form of an entire spectrum that we can judge that a certain percentage of them on the average must have been from background or overlaps. There is a technique that has enjoyed brief notoriety from time to time that increases the contrast in X-ray maps by "suppressing background". The method used is simply to reject or erase all dots which do not have another dot within a

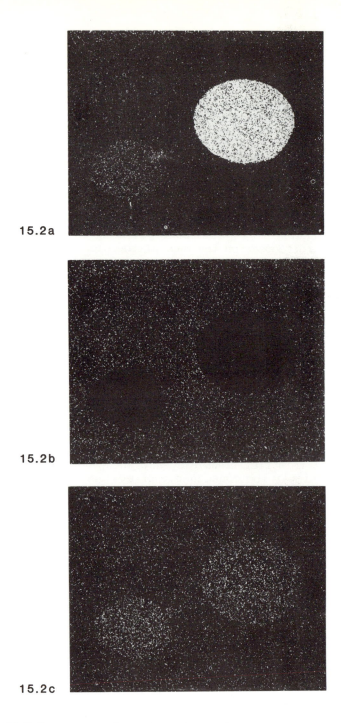

15.2a

15.2b

15.2c

Figure 15.2 X-ray maps from Cu and Zn wires in graphite. a) a 'normal' map for Cu (note the slight increase in dots in the region of the Zn wire); b) contrast reversal at high count rate; c) background intensity map showing atomic number effects.

certain distance. This is straightforwardly accomplished with a timer that permits the pulse to go to the display amplifier only when it follows another pulse within a certain time period (integrating over several pulses to get an average local count rate does not change the underlying principle, and further degrades the image resolution). Since time is proportional to distance for the scanning beam, this translates to distance between dots or to dot density. Removing dots in the low density regions does nothing to correct the various wrong reasons that the intensity might have been low there, and in fact will only distinguish between two areas if the average dot spacing is rather different in the areas to begin with. In other words, the method will accomplish electronically nothing more than you could yourself with a black felt tipped pen, to eliminate the dots in regions where you in your own judgement feel that the dots represent a lower, or perhaps a background level of intensity.

Ratemeters

Most of the limitations of X-ray maps have in one way or another to do with counting statistics. There just are not enough counts per unit of the image (per pixel) to define the count rate very precisely. If we could spend enough time, this would improve. One way to improve the statistics is to spend the several minutes (at least) that it takes to get a marginal quality X-ray map, covering perhaps several hundred by several hundred points on the image, scanning slowly along a single line and counting. This improves the statistics by a factor of several hundred, and so smaller variations in intensity can be seen. The way this is usually done is with a ratemeter connected to the pulse output which would otherwise go to the display. A simple analog ratemeter converts counts per unit time to a continuously varying voltage by allowing the pulses to charge up a capacitor which is continuously being discharged through a resistor. The R–C time constant defines the speed with which the ratemeter can respond to changes in count rate. The voltage is used to deflect the beam in the vertical direction on the display CRT while the horizontal position again corresponds to position on the sample. The result is a simple line profile of intensity along the scan line, in which the statistical fluctuations are evident and "real" variations can be seen if they are statistically significant. Figure 15.3 shows an example in which the profile is photographically superimposed on the SEM image.

A more recent development is the "digital linescan" or digital ratemeter. Its output may appear on either the CRT display of the microscope or of the multichannel analyzer or computer. In either case, it is produced by counting the number of X-rays stored in the region of interest for a short period of time, as the beam scans across a small region perhaps a hundredth or less of the image width. The number of counts is then stored or recorded, and the process repeated for another unit of time. If the number of counts is simply accumulated in one channel after another of the MCA (after all, we are monitoring only one element so we do not need the MCA memory for spectrum storage), the technique is identical to a method used for monitoring time decay of radioactive sources, and many multichannel analyzers, originally designed for the nuclear physics market, already incorporated so-called "multichannel scaling" capability with selectable integration times. Alternately, the number of counts can be accumulated in a register that is used with a digital to analog converter (DAC) to produce a voltage that can be used to deflect the CRT, just as the analog ratemeter did. These scans are slightly different in appearance, because of the stepwise changes in voltage, but the information content is similar. With either of these digital counting techniques, the data may be "dead time corrected" to compensate for changes in the overall system dead time, which represents a further improvement over analog or dot mapping techniques (Russ,1979).

15.3

Figure 15.3 X-ray line scan for Cu superimposed on the image from Figure 15.2

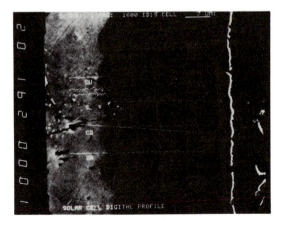

15.4

Figure 15.4 Digital presentation of elemental intensity profiles across a junction in a solar cell (the angled line shows the path of the trace) (photo courtesy Tracor Northern).

188

A comparatively recent extension of this technique is to allow the simultaneous counting of several different energy regions, which may correspond to different elements or to characteristic lines and background regions. If several elements are monitored, it allows the direct display of elemental intensities and their ratios either on the microscope CRT (Figure 15.4) or on the MCA screen, in the latter case with color displays added for clarity and/or visual appeal. While convenient and useful, these should not be mistaken for quantitative information, which as we have seen they are not (just the raw total intensity of peak plus background), simply because the word "digital" has been attached to them. They represent at best semi-quantitative indications of relative concentration changes, and the extensive computation required to extract quantitative information is scarcely justified by the poor statistics of the data.

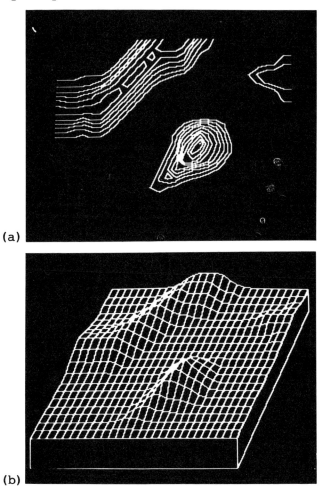

(a)

(b)

Figure 15.5 Presentation of two-dimensional maps of elemental intensity, P/B ratios, or elemental combinations or ratios, can be accomplished with grey scale or psuedo color contour plots (a), or isometric views (b).

189

plays of ratios of intensities (between elements, or peak to
ound ratios as was described in an earlier chapter) can be
ly produced from the stored elemental intensity scans. Figure
shows several examples of such profiles, across a complex
ic. Presentation of two-dimensional information with either grey
e, contour map, or isometric views of the data (Figure 15.5) is
) very useful for interpretation.

Problems with Surface Topography

Until now we have quietly assumed that the sample was flat
surfaced, as it would have been of necessity in the conventional WD
microprobe where these techniques began. However, in the SEM the
sample is more likely than not something with a rough surface, like a
fracture or particle dispersion, and in the STEM it may be a thin
section whose density varies markedly from place to place. In either
case, the simple plot or map of intensity versus position which we
have been counting on to reveal information about the distribution of
an element suddenly becomes distorted by confusing information. For
instance, in the case of a rough surface, the generated intensity of
characteristic X-rays will vary with tilt because of backscattering
effects, and the emitted fraction will rise or fall with absorption
because of the changing X-ray takeoff angle. These effects,
particularly the latter, can completely dominate the image. Large dark
areas may show not regions where the element is low in concentration,
but rather places that are shadowed from the detector position. Bright
regions may likewise be regions tilted sharply toward the detector so
that little absorption occurs. X-ray maps on rough surfaces are as
close to meaningless as any data you can take with a SEM, yet they are
routinely recorded and even published.

So too with thin sections: the variation in average atomic number
and mass thickness from point to point produces a corresponding change
in the continuum intensity that is generated. Since this is counted or
imaged in addition to the characteristic X-rays, it usually dominates
the entire image. Photographs using X-ray maps to show the
localization of some minor element, such as Ca or K, in heavily
stained cell membranes are most likely not calcium maps at all, but
rather density-atomic number maps . There may indeed be calcium there,
but the X-ray maps or line scans don't prove it.

A partial solution to this problem lies in the use of multiple
line scans to record the intensities from the elemental window
(characteristic plus background) and a nearby background window at the
same time. The ratio of peak to background, you will recall, has been
used as a quantitative indicator of concentration in thin sections and
a semiquantitative indicator of concentration in bulk samples with
rough surfaces. Just what we need, right? Well, sometimes it is. Don't
overlook the fact that the proportionality of peak to background ratio
with concentration is true only in a constant matrix, and more
specifically in a matrix whose average atomic number does not change.
For the thin section example, we could indeed use a peak to background
line scan to show concentration changes for an element if the matrix
average atomic number remained everywhere uniform. However, when heavy
element stains are used in biological tissue the average atomic number

may change by several hundred percent. Nor is the effect restricted to organic samples; heavy or light element precipitates in metal thin foils often have markedly different average atomic numbers than the matrix, so that plots of P/B ratio versus position may vary even in the opposite direction to elemental composition changes.

The situation is no better for rough surfaces. It is true that to a fair approximation the use of peak to background ratio will reduce greatly the dependence of the line profile on surface orientation, because the absorption of the background is at least approximately the same as for the characteristic lines. Of course, areas that are in complete shadow will still give no signal at all, but there is nothing we can do about that except change the sample orientation. But while getting a flat, unchanging line scan of elemental peak to background across a rough surface may be taken as an indication that the elemental concentration does not change, the converse is not true. Changes in the concentration of second elements can alter the average atomic number to produce varying P/B ratios quite independent, and often opposite to (because if one element's concentration goes up, the other will have to go down) the variation in concentration of the element of interest. Very confusing, then, is perhaps the most generous thing we can say about line scans or maps of peak to background ratio. (Maps are made by repeating the line scan along a series of lines across the surface. The data are usually stored up in the computer memory for display of several elements, logical combinations of elements, and pretty color pictures afterwards).

Chapter 16

Heterogeneous Samples

The preceding chapter attempted to describe the inadequacy of X-ray maps, and even line scans or processed intensity images, for characterizing heterogeneous samples. Yet it is these kinds of samples that represent the greatest area of application of electron beam instruments, which are capable of resolving variations in composition down to about 1 micrometer in bulk materials and to much smaller dimensions in thin foils. Let us now explore ways to determine and describe such inhomogeneous samples.

The point has been made several times (although its importance justifies repetition) that you cannot obtain an "average" composition of a heterogeneous sample by acquiring a spectrum while scanning the beam over a large area of the surface, and then treating the spectrum by the normal procedures for quantitative analysis. In the extreme case, if the sample consisted of abutting pure element islands, the k-ratio of intensity divided by pure intensity for each element in the spectrum would be just the area fraction represented by each element on the scanned surface, with no interelement effects at all. The quantitative correction program makes the implicit assumption that the matrix through which the X-rays pass has a composition given by the combination of all the elements present. Hence, in this case the Z-A-F corrections would be applied erroneously.

However, consider the analysis of a multiphase sample in which a representative spectrum from each single phase can be obtained. The same principle then suggests that the spectrum obtained by scanning over a large area of the specimen is simply a linear combination of the individual phase spectra. The fitting coefficients, which can be determined by the standard fitting software in many X-ray systems, are the area fractions of each phase. From the principles of stereology, the area fraction equals the volume fraction of each phase, provided the surface examined is a random one. This is then a useful and simple tool to determine volume fractions of phases in complex materials.

Furthermore, the individual phase compositions can be determined by conventional means from their individual spectra. These can then be combined to obtain the correct average or bulk composition of the sample by

$$C_j = \Sigma \; C_{ij} \; V_i \; \rho_i$$

where the subscript j corresponds to each element in the material, and i to each phase, and the density ρ of each phase must be estimated by the user, based perhaps on the composition.

The average composition determined in this way will in general be quite different from that obtained in error from applying correction procedures to the area-scanned spectrum. Table 16.1 shows an example of this for a three-phase ceramic, and applications to metal alloy composites have appeared in the literature.

Table 16.1 Intensity and concentration results for multiphase ceramic.

phases:	Hollandite	Perovskite	Zirconolite
Intensity (k-ratio=measured intensity / pure intensity)			
Al	0.0371		
Ca		0.3254	0.1079
Ti	0.3872	0.3269	0.1920
Zr			0.4288
Ba	0.2105		
Computed phase concentration (balance oxygen)			
Al	7.35		
Ca		29.45	11.79
Ti	39.15	35.25	21.760
Zr			40.07
Ba	18.71		

Bulk composition based on:	Volume Fraction	"Average spectrum"
Al	3.7	3.2
Ca	10.3	10.5
Ti	33.8	35.3
Zr	10.0	13.1
Ba	17.4	13.3

Scanning in a continuous raster over a large area is not the most efficient way to sample the specimen. Many SEM's, STEM's and microprobes provide means for external control of the beam position. This may be done either directly (a computer issues an X-Y address and the beam goes to that point) or indirectly (the beam moves stepwise in a raster pattern, with the step advance controlled by the computer). The first method, in particular, allows some very flexible approaches to characterizing heterogeneous samples. For instance (Russ, 1981c), the electron beam may be positioned briefly on each of very many points on the sample (perhaps using control of stage motion to increase the area examined). At each point, the intensities for each of several major elements in the sample are counted. The analysis time is very short, so as to permit sampling many points. These intensities have not the statistical precision to justify any quantitative corrections, nor even to attempt to separate background from peak. Instead, the observed intensity for each element, divided by the total of all intensities, is used as a rough indicator of elemental amount and used to plot a point, in n-dimensional space (where n is the number of elements).

The intensity ratio is not a measure of concentration, because of the different efficiency of generating X-rays for the different elements present. However, it is at least roughly proportional to concentration and allows separating data points from different concentration levels of the element. The use of the ratio also normalizes out any instrument instabilities and much of the effect of minor surface irregularities or changes in electron backscattering because of atomic number differences.

After many such points have been measured and plotted, clusters of points will be present representing the various phases present. The scatter of points within each cluster is due primarily to counting statistics, but still the center point, the average of each cluster, represents a good estimate of the mean intensity for the phase. These values can be used in a conventional quantitative correction scheme to determine the individual phase composition. Similarly, the number of points in each cluster represents the area fraction (and hence the volume fraction) of each phase.

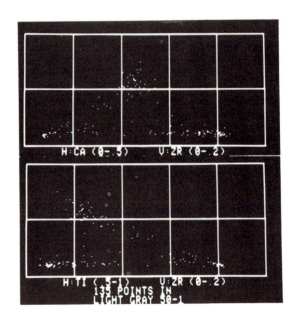

Figure 16.1 Plots of normalized intensity ratios for Al, Ti, Zr and Ca from multipoint analysis in multiphase ceramic: top) projected in the Ca/Zr plane; bottom) projected in the Zr/Ti plane

Notice that this method works even if it is not known beforehand what phases are present, nor even how many there may be. The n-dimensional plots of points are hard to visualize or examine, but two-dimensional projections (onto two element planes) can be used to identify the cluster locations visually. Alternately, computer pattern recognition techniques can be employed to find them. Figure 16.1 shows two-element projections of these intensity plots for a complex

194

ceramic. The additional points lying along lines between clusters result when the electron beam strikes a boundary between phases, and this increases in probability when the grain size is small (as in this example).

One drawback to this method, beyond the need for computer control of the electron beam, is the great amount of time it takes to measure enough points to detect the presence of minor phases. This is a strictly statistical problem, because the chances of striking a minor phase are small, and it requires some minimum number of points within a region of n-space to be identified as a cluster. The problem is worse when fine grain size or other variables produce a background of confusing points above which the cluster must appear. Even with counting times as short as 10 or 20 seconds per point, thousands of points require a lot of instrument time. In this situation it is natural to turn to the backscattered electron signal, which also contains some compositional information. Because there are 4 – 5 orders of magnitude more backscattered electrons than X-rays emitted from the sample, the statistical variation is small and very short measurements yield acceptable precision.

The fraction of electrons that backscatter from the sample is a function of average atomic number. Even if the backscattered electron detector intercepts only a small fraction of the total, and the fraction varies with specimen orientation (annular detectors are best for this purpose, but by no means essential), the backscattered electron signal is still generally able to show variations in image brightness that reflect differences in average atomic number. If the phases in the sample have discernible atomic number differences (as small as 0.1 Z with a favorable detector position), then phase distinction can be accomplished in this way.

If the voltage from the backscattered electron detector is digitized and read by the computer, either while the beam is being controlled or even during a continuous raster scan (digitization takes only 10-50 microseconds, typically), then histograms of the number of points of the sample surface with each level of signal brightness can be accumulated (Hare,1982). In these histograms, or spectra, the horizontal axis is atomic number and the peaks represent each of the phases present. The size of each peak gives the area (and hence volume) fraction of the phase. Location of a representative analysis point using the backscattered electron signal as a guide can then be used to obtain the composition of each phase in the conventional way. Figure 16.2 shows a backscattered electron image from a multiphase metal alloy, with its corresponding brightness histogram in which the phase peaks are evident.

Phase analysis using backscattered electron images from bulk materials, or specimen current, and using transmitted electron images from thin foils, requires some discussion of stereology. This technique, or family of them, allows three-dimensional parameters to be determined from measurements on two-dimensional images. For instance, we have seen that the volume fraction of a phase in a solid can be determined by measuring the area fraction which it represents on a surface through the material. Likewise, measurements on the surface image can reveal the amount of grain boundary surface area, mean free spacing between particles, grain size, and so on. These techniques are well known and widely applied in light microscopy, and will be increasingly used with electron microscopes, taking advantage of the computer inside many X-ray spectrometer systems.

(a)

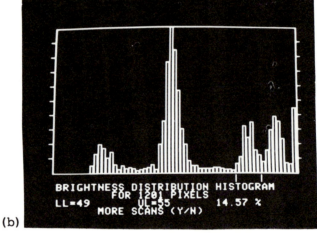

(b)

Figure 16.2 (a) Backscattered electron image and (b) brightness histogram

The measurement parameters described so far represent averages over the surface area. It is also possible to derive information from measurements on the individual features in the image. For instance, consider the case of particles embedded in a matrix. The surface area cuts through particles showing something about their size. But the image does not show the maximum diameter of the particles in this case, and a correction procedure developed by deHoff and Rhines must be used to estimate the actual size of the particles from the sizes of the intersections seen in the features. The correction coefficients are based on assumed shape factors for the features, and appropriate values for these can be determined from the image. For instance, if a shape based on an ellipsoid of revolution is employed as a model for the particles, then it can either be a prolate ellipsoid (in the extreme this can model needles or fibers) or an oblate one (tending in the limit to plates or disks). But for either type of ellipsoid, the two-dimensional sections seen will be ellipses. The only way to distinguish which shape is present is to examine a statistically meaningful number of shapes. If the minor diameter of the most un-equiaxed sections is close in size to the size of the most equiaxed ones, then the features are needles, or prolate ellipsoids. Conversely, for oblate ellipsoids, the major diameter of the most un-equiaxed sections will be close in size to the most equiaxed ones.

Such consideration go somewhat beyond the intended scope of this text, dealing not only with the compositional analysis of samples, but also with their structural analysis. The principal justification for the diversion is that stereology is a valuable adjunct to microanalysis, necessary to understand or characterize heterogeneous materials, and that it may often be possible to implement the programs for image measurement in the same computer as that used for X-ray data processing. The interested reader is referred to texts by Underwood and Weibel for further information.

A quite different sort of problem (the pun will eventually become clear) is to classify discrete particles, often sitting on a substrate (as particles on a filter), based on their composition. Some use of the backscattered electron image to locate the features, perhaps automatically, will speed the process up, but it is still the X-ray intensities which shall be necessary to estimate the composition and carry out the classification. Several schemes for such sorting have been implemented, each with certain advantages.

A later chapter discusses sorting in the context of alloy identification using X-ray fluorescence analysis. In that case, while the sorting strategies are similar, it is possible to do some preprocessing of the data to convert intensities to concentrations. Because of the poor counting statistics from a few second analysis in the electron microscope, and in addition because for irregular samples such as particles on a substrate the relationship between intensity and concentration is complex, it is not usually appropriate to do anything with the intensities beyond normalizing them (I_i / total intensity) or using ratios (to background or to a single major element). Furthermore, because the particle surface is irregular, a choice must be made either to position the beam at some central location on the feature (typically the centroid of the projected image measured with the backscattered electron signal) or to scan over a small grid covering the particle (and perhaps some of the surrounding substrate, which we will assume contains none of the elements used for sorting). Neither method is without pitfalls. The first may in fact miss the particle altogether if it is concave or irregular, while the latter further reduces the actual number of counts obtained in a given time from the particle of interest.

The simplest approach to classification is shown schematically in Figure 16.3. The intensity values for the elements are compared to a list of ranges for each element in each previously established class (or type, or compound). If the intensity value lies inside the classification region, the particle is counted as being of that type, and perhaps its size or location is stored for future analysis. The establishment of appropriate limits for intensities for each class is the responsibility of the operator, who may be able to determine some by measurement on known specimens, but ultimately must define ranges based on judgement about the variability of intensities due to geometric shape factors, counting statistics, and possible composition variation. Then account must be taken of the overall arrangement of classes, to minimize the chance of overlap of different ranges, and to reduce the number of cases where the intensity values do not fall into any category and are marked 'unclassified'.

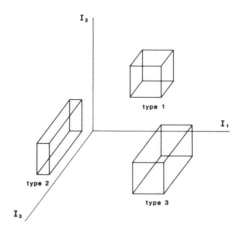

Figure 16.3 Classification boxes in intensity space.

While this method places a considerable burden on the analyst to establish proper categories, it is very fast in application. The sorting procedure is linear, that is the measured intensity value (usually a ratio) for the first element is compared in turn to each class. If a match is found, each successive element is compared to the class, and so on. The order of the classes and the elements can be optimized for speed, and even systems with hundreds of classes (various types of pollution particles , or minerals in coal, for instance) can be efficiently handled. However, adding an additional class to an existing set can have unexpected side effects on existing classes and the overall sorting sequence. The main limitation of the method is discriminating the miscellaneous or unclassified particles, and setting up proper categories in the first place.

A rather different approach to this kind of sorting begins with the same plot in n-dimensional space. However, the various classes are represented by either points (determined from representative samples, or perhaps computed by reverse application of a 'Z-A-F' quantitative correction program) or boxes or ellipsoids whose dimensions represent only the permitted variation in concentration. Each measurement point which is to be compared to these n-dimensional classes is plotted with finite dimensions which estimate its uncertainty. These can be adjusted for changes in counting time (and hence statistical precision) or other parameters.

Rarely will the plotted point (now actually an ellipsoid) lie within a class region. Sometimes it may intersect one, but most often the task is to find which region is nearest to it. This is done in a vector sense, by finding the straight line distance between the point and each class. The distance is then compared to the uncertainty of dimension of the point, and at some confidence level a cutoff is established for positive identification. Figure 16.4 shows the classification scheme. If the current analyzed feature is considered identified, it can be measured and counted as usual. If not, instead of being simply 'unclassified', the system will at least know what the most similar one (or few) of the classes were and that information can be saved to permit subsequent adjustment of the classes themselves (Karcich,1981). Alternatively, a longer analysis time or a more extensive pattern of points to cover the feature can automatically be employed to improve the quality of the measurement point.

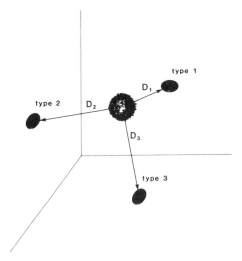

Figure 16.4 Classification scheme to find the nearest category.

The advantage of this approach is greater flexibility to deal with unclassified particles, including a learning ability based on the location of n-space points actually measured. It also reports matches or 'most-like' assignments reliably with much less initial user setup. Since all classes are always examined, file order is unimportant. It is much easier to set up and maintain a set of classes. On the other hand, a greater computational overhead is required, and a slightly greater time is needed to carry out the procedure. Nevertheless, as more computing capability is supplied in systems, this seems a better approach for the general application of particle classification.

Chapter 17

Less Common Methods

Much of the descriptive material and diagrams in the preceding chapters dealing with X-ray microanalysis (ie. excitation with a focussed electron beam, in SEM, TEM or microprobe) has not really been intended to train users to go out and write their own Z-A-F quantitative analysis software for routine use. Rather, it has been my goal to build conceptual understanding that will assist the user in choosing the proper analytical procedure to answer his or her questions about the sample, or indeed to pose meaningful questions for analysis. Armed with a mental picture of the electrons losing energy in the sample, in a roughly spherical region under the surface, whose size and shape vary in predictable ways with composition and voltage, the effects of sample geometry (both local surface irregularities and arrangement with respect to the beam and detector) become less mysterious. The role of X-ray absorption (and consequent secondary fluorescence) is more obvious when the fundamental behavior of atoms, being most strongly absorbed when they are near (and just above) the absorption edge of another element, is understood. And so on.

Armed with such a conceptual understanding, even without the details of the equations, it is easier to ask intelligent questions of the computer programs embedded in most modern analytical systems, and to test the answers for reasonableness. As a self test, try to approximately or qualitatively answer the following questions, and then figure out how you can use your computer system to get its answers:

1) Using a 12 keV. electron beam, perpendicularly incident on a flat surface of iron pyrite (FeS_2), with an X-ray takeoff angle of 20 deg., the measured intensities are 2300 cps for S K-α and 1000 for Fe K-α (above background). What will happen to the peak ratios if a) the beam voltage is raised to 25 kV.; b) the surface is tilted 60 deg. giving a takeoff angle of 80 deg.; c) both of the above. Try to identify the major factors operating in each case and estimate their relative importance.

2) Two specimens of the same 20%Cr-10%Ni stainless steel are excited with a 30 keV electron beam. One is a classic bulk sample and the other an ideal thin foil. Will the peak to background ratios for Cr and Ni be the same, greater, or less for the foil?

3) Three samples contain the same few percent of calcium; one is a primarily titanium oxide ceramic, the second is a silicate mineral, and the third an organic matrix. Which would you expect to give the greatest (and which the least) calcium X-ray intensity using a 15 keV electron beam. Why?

Once the reasoning paths used in these examples become natural, you can apply the same thought processes to the routine samples you encounter.

Handling Non-standard Situations

When real analytical problems are encountered, they rarely prove to be the classic flat-surfaced, homogeneous type amenable to Z-A-F correction. Some of the methods already described, such as the use of peak to background ratios to approximately correct for surface irregularities, arose in response to the need to handle less than perfect "real" samples, and were based largely on physical reasoning from the concepts that have been introduced here. One good example of a common problem might occur as follows: A qualitative analyses of a surface of sheet steel shows a small peak corresponding to sulfur. Is it a) sulfur in the steel, distributed homogeneously, b) a surface contaminant such as a sulfate primer, incompletely removed, c) subsurface inclusions (sulfides) that do not appear as defects on the surface?

Cross sectioning the sheet is one way to proceed, but is time consuming and may not hit the proper area to settle the question. From our simple model of the size of the excited volume, we know that we can change the depth of penetration by changing the accelerating voltage. We have also seen that the size of the sulfur peak, relative to that of the iron peak, should vary in a predictable way with voltage if the element is homogeneously distributed. The only terms in the equation for relative intensity that depend on accelerating voltage are the effective current R, excitation cross section (roughly $(U-1)^{5/3}$), and the absorption correction $F(\chi)$. Each of these can be straightforwardly calculated for another voltage, higher or lower than the original one. Some systems will in fact perform a semiquantitative analysis by dividing the measured elemental intensities for the sulfur and iron by these calculated relative intensities to report approximate concentration ratios.

If the predicted change in intensity is about the same as that observed (or if the semiquantitative concentration ratios are similar for the different accelerating voltages) then we should conclude that the sulfur is reasonably homogeneously distributed. On the other hand, if the sulfur intensity at the higher voltage is lower than we would expect, then the sulfur is predominantly at the surface, and the higher voltage electrons are penetrating through it and exciting more of the iron, relative to the sulfur. Conversely, a sulfur intensity higher than we would expect at the higher voltage suggests that the sulfur is below the surface. If that is all the information required, then our task has been easily acomplished without extensive sample preparation, measurement of standards, or lengthy computer programs. Even if more questions are raised, at least we know enough to proceed with the best experiment to answer them.

Coating thickness measurements: Another common problem dealing with the depth dimension in samples is that of measuring coating thickness. We have already seen that in the general case of unknown composition and thickness of the coating, Monte-Carlo calculations can sometimes be used to construct a family of calibration curves relating intensities to the elemental concentrations and thicknesses. If the problem is simply to measure thickness, there may be an easier way to proceed. Given the availability of one, or a few standard samples, we can construct a calibration curve in several ways.

For example, the ratio of the intensity from an element in the coating to the intensity from an element in the substrate I_c/I_s should be a very sensitive measure of thickness. As thickness increases, in the range where the electron beam penetrates through the coating into the substrate (which we can predict using the electron range nomogram introduced in Chapter 4), the intensity from the coating element will increase and that from the substrate will decrease, so that the ratio is a sensitive measure of thickness as indicated in Figure 17.1. Because all of the processes (electron stopping, X-ray absorption) are roughly exponential with distance, it is not unusual to obtain log-linear plots for these calibration curves. Knowing that, we might construct a curve with a minimal number of standard samples.

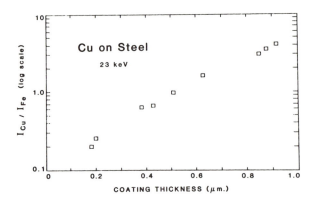

Figure 17.1 Typical calibration curve for intensity ratio vs. coating thickness

Even with just a single standard, a series of measurements at varying tilt angles (using the cosine of the angle to estimate the foreshortening of the depth of electron penetration) may enable the construction of a calibration curve. Finally, even with no standards at all, if the accelerating voltage can be varied and the intensity ratio plotted, behavior such as that shown in Figure 17.2 can often be observed. The "break point" in the curve corresponds to the voltage at which the depth of electron penetration corresponds to the coating thickness. This can then be obtained from the electron range nomogram. Again, tilting the sample can sometimes bring this distance into the easily measured range, although the lateral spread of the electrons in the sample may influence the results. Variants of this technique in which only the substrate contains elements which emit measurable X-rays can be used to estimate the thickness of organic coatings, as well.

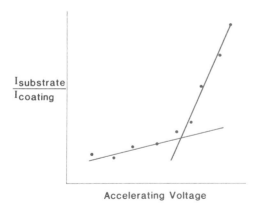

$\dfrac{I_{\text{substrate}}}{I_{\text{coating}}}$

Accelerating Voltage

Figure 17.2 Schematic diagram of intensity ratio plot versus accelerating voltage.

The case of the unknown takeoff angle: Frequently in the analysis of fractured metal or ceramic parts, it is necessary to determine quantitatively the composition at a location (perhaps the origin of the fracture) which is locally flat, over a distance of a few micrometers, but whose orientation with respect to the electron beam and detector is not known nor readily measured. In an earlier chapter the technique of stereoscopy (parallax measurement) was introduced, but this is time consuming at best, and depending on the stage motions available in the microscope may be impossible to carry out. This turns out to be another case where changing the electron accelerating voltage can permit calculating quantitative results.

If two analyses of the same location, with the same surface orientation and of course the same composition, are carried out at different (and carefully measured) voltages, they should give the same analytical result. Of course, there are inevitable errors, not least of which may be the purely statistical ones in the intensity values. Nevertheless, we would naturally expect the two sets of data to agree most closely when we performed the calculations with the correct X-ray takeoff angle value in the calculations. The computational procedure is thus to guess at a takeoff angle, compute concentrations (using a normalizing, no-standards Z-A-F routine since we will surely not have standard intensities for the general case of unknown takeoff angles) for each set of intensity data, use the differences between them as a test parameter, and vary the assumed takeoff angle until the difference is minimized (Russ, 1980a).

Computer generated curves such as those in Figure 17.3 suggest that for complex multi-element samples the pattern of peak intensities is a unique function of takeoff angle (in this case varied by varying the surface tilt). In this example, conditions can be found at which any element has the smallest or largest peak, but no two angles have the same set of intensity ratios.

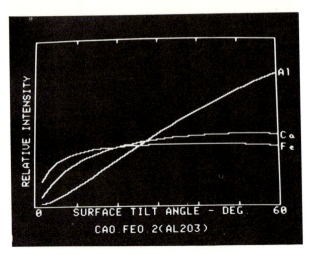

Figure 17.3 Calculated (by reverse application of ZAF equations) intensities for three elements in a complex oxide. The detector position relative to the sample surface is at the same height as the sample and 60 degrees in azimuth from the direction of tilt.

In the particular procedure we use, the first assumed angle is 15 degrees (in the range where intensities change rather rapidly with takeoff angle due to substantial absorption effects). After concentration values are computed for this angle, 20 degrees is chosen and the process repeated. At each angle, the sum of squares of the differences (SSD) in concentrations for all elements present is computed. This becomes the parameter to be minimized. Depending on whether the 15 or 20 degree data gave a lower SSD, the next angle chosen is either 10 or 30 degrees. Once three points have been calculated, a parabola is fit through the SSD data to predict the angle at which the minimum will be reached. The new angle thus chosen is used as the next guess and the poorest (largest SSD) angle is forgotten.

The process continues until convergence is reached, defined as a change in angle or composition less than some preset amount (we use 0.2 degrees or 1% relative change in SSD). The average of the two sets of concentration data then give the best estimate of true concentration of the sample, and the angle value can be examined as a check on the reasonableness of the estimate. Figure 17.4 shows the concentration and SSD curves for two examples of this technique. For the olivine, the "true" takeoff angle was 13 degrees. Note the sharp minumum in the SSD curve, which is not present for the stainless steel case (for which the actual takeoff angle was 23 degrees). This is because in the latter case, the higher energy X-rays are absorbed less in the matrix and hence the corrections and intensities change less rapidly with the assumed takeoff angle. However, this means that the resulting concentration values are also less sensitive to the exact choice of angle, and so the computed results are still quite good. Tables 17.1 and 17.2 show the comparison of these results with the quantitative results from conventional Z–A–F calculations using an independently measured takeoff angle.

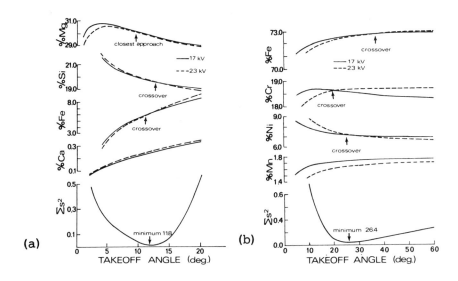

Figure 17.4 Curves of calculated concentration and SSD value versus assumed takeoff angle: a) olivine mineral, b) stainless steel.

Table 17.1 Olivine concentrations*

	nominal	conv. ZAF	angle iteration
MgO	48.5	48.52±.20	49.43±.30
SiO$_2$	40.8	42.65±.19	42.28±.56
CaO	0.21	0.37±.04	0.34±.02
FeO	9.8	8.47±.15	7.95±.81

* indicated range based on ten separate analyses for conventional ZAF and 25 combinations of values for angle iteration.

Table 17.2 Stainless steel concentrations*

	nominal	conv. ZAF	angle iteration
Cr	18.37	18.66±.20	18.46±.13
Mn	1.71	1.77±.13	1.66±.08
Fe	71.1	72.35±.22	72.39±.28
Ni	7.40	7.23±.15	7.50±.38

* indicated range based on eight separate analyses for conventional ZAF and 12 combinations of values for angle iteration.

Although as was shown in the chapter on errors, the sensitivity of the angle iteration method to errors in voltage measurement is greater than the conventional ZAF method, that parameter is easier to determine accurately than takeoff angle. Consequently, this method seems to be particularly useful for SEM microanalysis, in which samples are often encountered whose local orientation cannot be readily measured.

Using the backscattered electron signal: It often happens in microanalysis, especially of mineral, oxide or carbide samples, that at least one element must be determined either by stoichiometry or by difference using conventional ZAF programs. "By difference" means that all of the analyzable elements are computed and at each iteration the difference from 100% is assigned arbitrarily to the remaining element (eg. oxygen). "By stoichiometry" means that a fixed atomic ratio of the unanalyzed element is entered by the user (from independent knowledge) and used to compute the amount of the element present throughout the iterative calculation based on the analyzed elements' concentrations. In both cases, the user is called upon for judgement as to which element is missing. Furthermore, the "by difference" method requires standardization of the analysis, rather than the often more convenient normalized no-standards method.

Another way to determine the missing element, and normalize the concentrations without requiring a standard X-ray measurement, is to use the backscattered electron signal (Russ, 1977b). In Chapter 4, the monotonic relationship between atomic number and the backscattered electron fraction was noted. It is quite difficult to design a backscattered electron detector for most electron microscopes that collects the total of all the backscattered electrons, but for a given sample geometry it is reasonable to expect a fixed position detector to detect a reproducible fraction of the total signal. By prior calibration with a series of even a few pure metals covering the range of interest (generally rather low atomic numbers since the whole point of this exercise is to assist with unanalyzed low atomic number elements), a relationship between average atomic number and backscattered electron signal can be established for a given instrument.

Measurement of the backscattered signal, perhaps as a ratio to the total beam current collected in a simple Faraday cup, or to the signal from a known elemental standard (which can be the sample stage or holder itself), then gives a very rapid indication of the average atomic number of the unknown. This is much faster than an X-ray measurement to similar precision because of the much higher yield of backscattered electrons, and consequently better statistics. Since the backscattering coefficient η for a complex sample can be satisfactorily estimated as a linear combination of the coefficients for the individual elements

$$\eta_{matrix} = \Sigma \, (C_j \, \eta_j)$$

this provides the missing equation, along with the ZAF equations for the concentrations of the measured elements, to permit quantitative analysis. It can either indicate what element is missing in a conventional calculation, or permit a normalized no-standards analysis with the missing element specified.

Other uses for the backscattered electron signal are to provide analytical information directly, to locate features for X-ray microanalysis as described in a preceding chapter, or to estimate sample surface orientation by comparison of the backscattered electron signal reaching each of several detectors, or quadrants of an annular detector mounted above the sample on the microscope polepiece. Finally, it should be noted that the backscattered electron fraction will drop when the incident electron beam strikes the sample at a critical angle such that diffraction occurs (depending on the electron wavelength and the spacing of atomic planes) and the electrons penetrate much deeper into the sample. This will also alter the X-ray depth distribution and intensities, which can affect quantitative analysis (especially in thin sections where the electrons may pass through without generating any X-rays). These effects are not common, but the use of backscattered electrons to study crystallographic orientation in bulk samples is important in electronic materials, especially.

In some cases the energy distribution of the backscattered electrons can also be interpreted to gain analytical information about the sample. The effect of atomic number and surface orientation on the shape of the energy distribution of backscattered electrons was discussed in Chapter 4. With an energy dispersive detector, either a Si(Li) detector without an entrance window or a room temperature detector, these spectra can be measured directly and used to supplement X-ray measurements.

In the STEM, the electrons that penetrate through the sample lose energy by the various mechanisms that have been described, as well as by exciting plasmons in the sample lattice. A spectrum of the transmitted electron energies shows the absorption edges of the elements present in the sample, and can be used for elemental analysis. This spectrum is usually obtained with a magnetic sector spectrometer (although other designs with crossed fields have also been used). The so-called electron energy loss (EELS) spectrum is particularly useful for very low energies, and supplements the X-ray spectrometer nicely by giving information about the light elements. It is beyond the intended scope of this book to discuss the mechanics of EELS spectrometers, the interpretation and quantification of the spectra, or the other information contained therein (including structural information such as atomic nearest neighbor distances). Suffice it to say that a complete analytical electron microscope based on the STEM will incorporate this technique as well as X-ray analysis, and may well acquire data from both spectrometers into the same controlling computer (which may also be able to position the electron beam and control the microscope itself).

Chapter 18

X-ray Fluorescence – Fundamentals

Photon Excitation – X-ray Tubes

The use of X-ray or gamma ray photons to produce excited atoms, which in turn emit characteristic X-rays, is in many respects quite different from the use of electrons or other charged particles. The cross section for stopping (absorbing) the incoming photon is different, and this produces a different depth distribution of excitation, which in turn gives rise to different absorption and fluorescence effects for the generated X-rays. Also, the background in the measured spectrum is quite different because the photons cannot directly produce Bremsstrahlung. These effects, and some of the consequences for a mathematical model of the intensity-concentration relationship, will be dealt with in this and subsequent chapters.

First, it is useful to recall that in the great majority of instruments that perform X-ray fluorescence measurements, the excitation is produced by X-rays which are themselves generated by electrons. The conventional X-ray tube is simply a rather specialized electron beam, with an accelerating voltage of perhaps 10 to 60 keV (some systems go higher or lower, and dental or medical X-ray tubes and generators are often over 100 keV. to produce higher energy X-rays which will penetrate further). The target in the tube emits X-rays following the behavior which we have already described: the continuum or Bremsstrahlung radiation covers all energies up to the maximum voltage on the tube, and on top of that are the characteristic emission lines (K, L, and/or M) from the element(s) in the target. It is most common to select target elements based first and foremost on their electrical and thermal properties, because the tube may operate at power levels of several kilowatts (for instance, a voltage of 50 keV. and a current of 40 mA. is 2 kW.), much higher than the power level in even an electron microprobe (where a beam voltage of 30 keV. and current of 0.1 to 1 microamperes would be rather high, while in the SEM the beam current might be well under a nanoampere). The heat generated in these high power tubes is usually removed by a water circulation system, but even so, since the depth of penetration of the electrons is very small (compared to the thickness of the target or anode), the heat conductivity is important, and so is the ability of the target to maintain its integrity at somewhat elevated temperatures. Refractory metals are, however, not good thermal conductors. The usual arrangement is to make the anode assembly (unlike the electron beam instruments we have described before, the electrons are usually accelerated all the way to the target in the X-ray tube) out of copper, and apply a thin coating of the element to be used as a target.

The choice of a particular target element is sometimes based on the desireability of its particular characteristic X-rays for the excitation of elements of interest in the unknown. However, this would restrict the analytical possibilities greatly. Most of the excitation is in fact normally carried out by the continuum radiation from the tube, and to maximize this, the target element is often chosen with a high atomic number (since the total continuum intensity increases with atomic number). Tungsten (W, atomic number 74) is a common target material in X-ray fluorescence tubes, being high in atomic number and refractory. Molybdenum (Z=42) is also used, since it has characteristic lines in a very different part of the spectrum from tungsten.

Another element sometimes used for tube targets is chromium (Z=24). It would seem at first that this would be a very poor choice, since the chromium target will produce only a third as much continuum for a given beam current and power level than the tungsten target. But in order to get out of the X-ray tube to reach the sample, the X-rays must pass through a beryllium window which isolates the comparatively high and clean vacuuum of the sealed X-ray tube from the sample chamber. This window is placed very close to the target since all of the distances are kept short to direct the maximum intensity of photons from the tube to the sample. Consequently, it is struck by many of the backscattered electrons from the target, which heat the window. Beryllium is used, of course, because of its good mechanical strength and high transmission of the low energy X-rays needed to excite the light elements. But beryllium is not a good conductor of heat, and in order to work with the high atomic number target materials (which have high backscatter coefficients, we recall), the thickness of the window must be rather great, perhaps 125 micrometers or more. This causes severe absorption of the low energy photons. The use of a chromium anode (actually a plating of Cr on copper, of course) reduces the backscattering permitting a much thinner window. Also, the chromium characteristic line at 5.4 keV is somewhat useful for exciting lighter elements, and the metal itself will withstand fairly high operating temperatures. Consequently, chromium tubes are often used for the analysis of lower atomic number elements.

With the advent of energy dispersive detectors, a somewhat different set of criteria has arisen for tube parameters. The count rate capability of these systems is vastly inferior to the wavelength dispersive spectrometer, since the pulse width through the main amplifier must be several tens of microseconds (full width), to permit low-noise measurement, rather than a fraction of a microsecond for the pulse from the detector used in the wavelength detector (where discrimination between X-rays of different energies or wavelengths, from different elements, has been achieved by the diffracting crystal, and only counting is required of the electronics). Consequently, it is not necessary to use such high power tubes and generators. This reduces the cost and complexity of both the tube and generator greatly, usually eliminates the need for water cooling (although an oil filled bath may still be used to conduct heat from the anode to a larger cooling surface), and of course reduces or eliminates the problem of heating the beryllium window.

Consequently, it is practical to build low power X-ray tubes for energy dispersive systems (operating at a few tens of Watts rather than several kW.) which have quite thin beryllium windows (nearly as thin as the 7-25 micrometer windows available for the detectors). These transmit much higher fractions of the low energy exciting radiation, and offer enhanced detection limits for light elements. Also, the choice of target element becomes much less restricted, and elements such as gold or silver, which would not be used in a high power tube because of the temperature problem, can be considered. This means that it is is possible to choose a target element based on its ability to produce continuum or characteristic X-rays, or both, to efficiently excite the elements of interest in a partciular class of samples. Anode materials such as gold and silver, which would not withstand the high temperatures encountered in conventional tubes, are both used. Several special tube configurations that provide modified spectral output to the sample will be discussed shortly. It is worth introducing here a type of tube that has become very common in energy dispersive systems, which incorporates a grid (which functions much like the grid in a triode vacuum tube used for signal amplification). Connected to appropriate circuitry in the main pulse amplifier, which was described in an earlier chapter, this can be used to turn the tube off when a pulse is being processed, so that the likelihood of pulse pileup is reduced and all of the generated pulses from the detector are counted. The total increase in stored counts for a given operating power level (averaged, of course, over many on/off cycles of the tube) can be 2.5-3 times, compared to a conventional tube (Jaklevic,1972).

It has also become possible to employ absorbing filters to modify the energy spectrum coming from the tube, to improve the analytical sensitivity, detection limits, or selectivity for particular elements of interest. We will return to this subject in a future chapter.

Absorption of the Generating X-rays

The radiation coming from the X-ray tube enters the sample and is absorbed following the usual exponential law:

$$I \, / \, I_o = e^{-\mu \rho t}$$

where I is the intensity (number of photons per second), I_o the original intensity, t the distance, ρ the density, and μ the mass absorption coefficient. Just as we saw before, the mass absorption coefficient for a complex material is the linear sum of the mass absorption coefficient of each element present for the energy of the X-rays, times the concentration (weight fraction) of the element. The mass absorption coefficient is greatest, we recall, and shows the greatest probability of the element absorbing the photon (and becoming excited) when the energy is just above the critical excitation energy (the "absorption edge energy") for a shell in the element. Between these edges the value of the mass absorption coefficient drops off exponentially with increasing energy. This behavior, which we have previously described, is shown as a reminder in Figure 18.1.

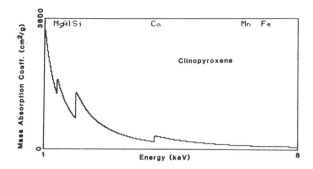

Figure 18.1 Variation of mass absorption coefficient vs. energy and wavelength for a complex mineral.

The excitation of each shell (K, L, M) is given by the "step height" or edge jump ratio. The fraction of the total absorption that goes into excitation of the particular shell whose edge we see (the K edge in the figure) is just $(S_K-1)/S_K$. For the L or M shell, the same expression would be used with the particular subshell (L I,II,III, M I...V) corresponding to the transition which generates the characteristic line whose intensity we will be measuring. This fraction will be multiplied by the fluorescence yield ω for the shell, and by the relative line intensity L for the α or other line, to give the total yield of X-rays.

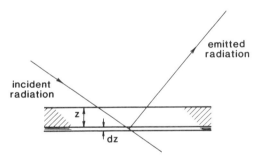

Figure 18.2 Diagram showing generation of radiation in a layer at depth z in the sample.

However, to get the total characteristic intensity generated for a particular line of a selected element in a real sample, it is necessary to consider the distribution of radiation in depth, as shown schematically in Figure 18.2. For each layer dz in the sample, and for each energy of exciting radiation E, we have:

$$J(E)\ C_i\ \mu_i(E)/\mu_{mtx}(E)\ e^{-\mu_{mtx}(E)\ \rho z/\sin\phi}\ (S_K-1)/S_K\ \omega_i\ L_i$$

where J(E) is the distribution of intensity as a function of energy coming to the sample from the X-ray tube, C is the concentration of the element (i) in the matrix, the angle ϕ is the angle at which the radiation enters the sample (90 degrees is p erpendicular to the

211

surface), and the other terms are as previously identified. The terms in this expression thus give the intensity entering the sample, the fraction penetrating to depth z, the fraction absorbed by element i, and the amount of generated intensity it produces. The fraction of this radiation which would be emitted would be further reduced by absorption leaving the sample:

$$e^{-\mu_{mtx}(E_i)z\rho/\sin\theta}$$

where E_i denotes the energy of the characteristic line for element i, and the angle θ is the X-ray takeoff angle from the surface. This absorption is called secondary absorption, to distinguish it from the primary absorption of the incident radiation, which was described by the first equation.

The total intensity from element i must be determined by a double integration of these terms, once over all depths in the sample, and once over all incident photon energies. Doing so, and pulling outside the integral the constant terms, gives the emitted intensity for element i as:

$$C_i(S_K-1)/S_K \;\omega_iL_i \;_{Eo}\!\int^{Ec} \mu_i(E)\;\rho J(E)dE$$

$$_0\!\int^{\infty} dz \;\exp(-z\rho(\mu_{mtx}(E)/\sin\phi + \mu_{mtx}(E_i/\sin\theta)))$$

which reduces to:
$$I_i = C_iq_i \;_{Eo}\!\int^{Ec} \frac{\mu_{mtx}(E)J(E)dE}{\mu_{mtx}(E)+A\mu_{mtx}(E_i)}$$

where the constant q includes various fundamental parameters specific to the element i and to the spectrometer, and A is a geometric constant equal to $\sin(\phi)/\sin(\theta)$.

Note that this equation expresses very simply the intensity we would expect to observe from a given concentration for element i in terms of parameters such as mass absorption coefficient, which we know how to look up or compute, and the tube spectrum, which we should be able to measure for a given system. However, the equation is deceptive. The μ_{mtx} terms (either a function of the exciting energy E over which the integrration is carried out, or evaluated at E_i the emission energy of the analyte element) which appear on the right side of the equation, inside the integral sign, are summations which include the concentration (weight fraction) of each element in the matrix. These concentrations are the unknowns, so if we have a set of measured intensities I from each element in the sample, this set of N simultaneous, nonlinear, integral equations would have to be solved to determine all of the concentrations. We shall hence find that this "fundamental parameters" equation (which is not yet quite complete) will serve more as a starting point for approximation, or as a conceptual model, than as an equation for solution.

One helpful consequence of this equation is seen if we plot against energy of the exciting photons, the emission spectrum of the

X-ray tube (which looks, of course, just like the spectrum from an electron excited specimen of a pure element, as we considered it before) and the matrix mass absorption coefficient. If the matrix is a single element, we have the situation shown in Figure 18.3.

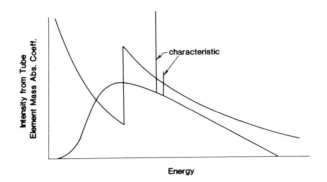

Figure 18.3 Superposition of spectrum from X-ray tube and elemental mass absorption coefficient, as a function of energy.

The integral need be carried out only from the absorption edge (below that, the incoming photons cannot excite the shell emitting the line we are measuring) to the voltage on the tube (above which there is no exciting radiation). Since the mass absorption coefficient drops off exponentially above the absorption edge, it is clear that it is the amount of exciting radiation J(E) near the absorption edge that dominates the integral, which is effectively the product of the two terms. This may be either the characteristic line(s) of the target element or simply the continuum, depending on the analyte element and the tube target material.

In a real sample, when there is a mixture of elements with different absorption edge energies, there is "competition" for the exciting radiation. Figure 18.4 shows schematically an example for analysis of chromium in a stainless steel. The presence of the iron and nickel, which make up perhaps 80% of the matrix, causes substantial absorption of the original tube spectrum. Consequently, the "effective" spectrum of photons available to excite the chromium is reduced. In actuality, the amount of the reduction is not so simple, and varies with depth in the sample. It is important to note, however, that the presence of matrix elements which absorb strongly just above the absorption edge of the analyte will significantly reduce the excitation of the element. This is called a "primary absorption" effect, and is one of the kinds of interelement corrections which we shall have to make. It is included, along with the "secondary absorption" of the emitted X-rays from the analyte by the matrix, in the equation above.

The third important interelement effect is that of secondary fluorescence. When the emitted characteristic X-rays of one element are absorbed by elements in the matrix, they cause additional excitation of those atoms beyond the number produced by direct primary absorption of the exciting X-rays. The situation is identical to that for electron excitation, which has already been discussed, except that because the depth of excitation is much greater for X-ray excitation than for electron, the amount of absorption and consequently the amount of secondary fluroescence is greater. Figure 18.5 shows schematically how secondary and even tertiary fluorescence can occur, again for the case of chromium in a stainless steel. The total intensity measured from the chromium is the sum of the direct excitation of chromium atoms by incoming primary photons, excitation due to chromium atoms absorbing primarily generated photons from the nickel and iron atoms in the matrix (secondary fluorescence), and excitation of chromium atoms by X-rays from iron atoms which were themselves excited by X-rays from the nickel atoms (tertiary fluorescence). Obviously, we could add more elements and propose the possibility of fourth or even higher order effects.

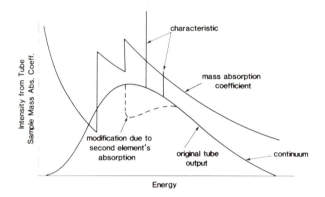

Figure 18.4 Schematic indication of the effective change in the tube spectrum available to excite one element due to the presence of another, shown by superposition of the tube spectrum and the sample mass absorption coefficient.

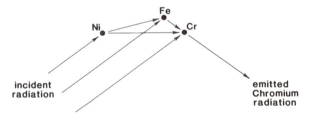

Figure 18.5 Secondary and tertiary fluorescence possibilities in a stainless steel.

Since we saw before, in the case of electron excitation, that the importance of secondary fluorescence is small unless the fluorescing element has an emission line quite close to the absorption edge of the analyte, we may expect that the same is true here. Indeed, even in a case such as stainless steels, which are a classic example of large secondary fluorescence corrections because the elements do have peaks

214

and edges that are close together, we find that the total magnitude of the terms is not all that great. For example, in a 10%Cr, 70%Fe, 20%Ni alloy, the relative contribution of each of these sources of excitation of the chromium is as listed below:

Cause of excitation	Percent
direct primary excitation	72.5
secondary fluor. by Fe	23.5
secondary fluor. by Ni	2.5
tertiary (Ni-Fe-Cr)	1.5

From this table, we see that even in this extreme case the total contribution of secondary fluorescence is only about a quarter of the emitted radiation. Furthermore, recall that the presence of the iron and nickel which increase the chromium intensity by fluorescence also reduced the primary intensity by competing for the incident radiation (a primary absorption effect). Normally, any secondary fluorescence correction will be at least partially offset by a primary absorption effect. The expression for the secondary fluorescence looks much like that which we saw before for electron excitation, since both describe the same process. We shall not write it out here simply because we will not find it necessary to solve for it explicitly. It can be thought of as entering the fundamental expression we saw above as an additive term in the numerator of the fraction inside the integral, since the same secondary absorption (the denominator terms) will be encountered by these X-rays as by the primary ones.

Also, since the table shows that the contribution of the secondary fluorescence drops off even as we go from iron, whose emission line at 6.4 keV. is just above the chromium excitation energy of 6.0 keV., to nickel with an emission line at 7.5 keV., and since the magnitude of the tertiary fluorescence term is extremely small, we see again that interelement fluorescence will be a comparatively unusual situation. We shall have to consider it only when elements have peaks and edges quite close together, and the higher order terms will in practically all cases be ignorable altogether.

If we do add the complete fluorescence terms to the equation, which makes them even more cumbersome, we clearly have a set of equations of little practical use. It would be possible, as was done above to predict the chromium intensity from the steel, to insert parameters and concentrations for a known composition, measure or approximate a tube spectrum, and carry out numerical integration in a computer to predict intensity from concentration. But since this is inherently much less useful than calculating concentrations from intensities, we must find a simpler way to proceed.

One way is to use the calculations of intensity as "virtual standards." By running one or a few physical standards, from which the appropriate proportionality factors can be determined for a given system and set of analysis conditions, a computer program is then able to compute intensities which would be measured from samples having enough different standard compositions to permit the regression methods to be described in the next chapter to be carried out. The advantage of this method is that as many physical standards as are available can be used, and then supplemented by computed standards to cover the inevitable "holes" in the ranges of the various elements present. Also, additional samples can be computed to more densely cover important regions. Finally, although this process is quite time consuming initially, the application of the computed regression curves to unknowns is acceptably fast.

This sort of program was first developed for rather large "mainframe" computers, but has now been adapted with very few compromises to laboratory minicomputers of the sort incorporated in many X-ray systems (Criss,1980). It requires either a measured and stored tube profile, or one computed using the sort of equations described in earlier chapters covering electron excitation.

The Effective Wavelength Model

In order to simplify the complete fundamental parameters equation which relates concentrations to intensities, which is shown in somewhat abbreviated form above and has been derived in detail by Sherman (1955,1959), Shiraiwa and Fujino (1966), Criss and Birks (1978), and others, several types of approximations have been introduced. Most will wait for the next chapter in which methods that lead to regression of linear simultaneous equations is discussed. The full theoretical treatment would require at least knowing the spectral distribution of the tube output $J(E)$, over which a numerical integration would then proceed. If the integration could be eliminated, the simultaneous equations might be workable.

A way to accomplish this comes from the mean-value theorem, which states that for any continuous function $Y(X)$ there must be some location X where Y takes on any chosen value between its limiting values, and in particular that there is some location at which it takes on the mean value. In other words, there must be some value c such that:

$$_a\int^b Y(X) \ dX = (b-a) \ Y(c)$$

As applied to our situation, it means that rather than having to integrate the product of $J(E)$ and $\mu(E)$, both of which are smooth continuous curves (at least between absorption edges) over the range from E_c to E_o, there must be some energy E in that range such that the product of J and μ at that energy give the same result as if the integration was carried out; in other words, the theorem states that there is an "effective" energy or wavelength (depending on how we plot the curves) such that if all of the exciting radiation had that energy, instead of a continuous distribution of energies, the excitation of the analyte element would be exactly the same.

It must not be supposed from this that we can readily find this energy, nor that it will be the same for the different elements in the matrix, not even that it will be the same for a particular element in different matrices. To make practical use of this method, we can measure the intensity for element i from a standard sample, and calculate what the tube intensity J would have had to have been for any "effective" energy E, for the mean value theorem to apply. Usually the guess at an effective energy is taken slightly above the absorption edge for the element, either a constant energy above it (eg. 0.3 keV.) or a constant multiple of it (eg. 1.05 times E_c), in order to compute the necessary mass absorption coefficients. These values can be justified on the grounds that we have already seen that the exponential fall-off in the value of μ with energy means that the most important exciting photons are those which lie just above the absorption edge. Also, experiment shows that these estimates of the effective energy for excitation work, when the exciting radiation has the broad distribution of typical Bremsstahlung (for monoenergetic excitation or strong characteristic lines in the spectrum, we will see later on that the effective excitation energy can be changed to the energy of the principal exciting radiation). Then the equation is solved for the effective tube intensity J^*.

If the standard sample is either a pure element (in which fluorescence effects are missing) or very similar to the unknowns to be analyzed later (which minimizes the error due to the guess at the effective E since small changes in mass absorption coefficients will cancel out as ratios), this method gives straightforwardly the "effective tube intensity", as well as the intensity which would be produced from the pure element. From these factors, obtained for each element of interest under a particular set of analysis conditions (tube voltage, principally, since time and current can be taken into account in expressing the intensity values), it is possible to compute concentrations for unknowns. The form of the equation is:

$$I_i = C_i q_i \, _{Ec}\!\int^{Eo} J^* \mu_i(E^*) \, (1+SF) \, / \, (\mu_{mtx}(E^*)+A\,\mu_{mtx}(E_i))$$

where SF is the secondary fluorescence term and all the other terms have the same meanings as previously introduced.

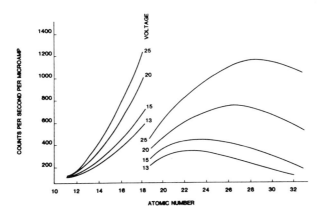

Figure 18.6 Curves of pure element intensity measured by X-ray fluorescence using various tube voltages (Ag target tubes).

217

Table 18.1 Analysis Results using the Effective Wavelength Model

A. NBS Type 638 Cement, Rh Tube, 13kV.

	Actual Conc (w/o)	Intensity (cts/sec)	Calculated Concentration stored curve	similar std*
Al_2O_3	4.5	12.18	2.96	4.05
SiO_2	21.4	158.62	19.25	22.86
SO_3	2.3	57.14	3.79	2.20
K_2O	0.59	9.56	0.57	0.63
CaO	62.1	1103.75	64.51	61.67
Fe_2O_3	3.58	62.08	3.92	3.59

* standard: type 635 cement

B. USGS AGV1 Rock, Rh Tube, 10 kV.

	Actual Conc (w/o)	Intensity (cts/sec)	Calculated Concentration stored curve	similar std*
Na_2O	4.07	15.19	3.05	4.31
MgO	0.76	17.57	1.25	1.13
Al_2O_3	15.40	338.37	16.63	15.31
SiO_2	69.11	2352.68	68.04	67.75
K_2O	4.51	358.21	3.36	4.42
CaO	1.94	177.13	1.96	1.80
TiO_2	0.50	40.98	0.46	0.49
Fe_2O_3	2.65	167.89	2.46	2.78

* standard: type GSP-1 rock

C. Type VII Stainless Steel, Rh Tube, 15 kV *

	Actual Conc (w/o)	Intesity (cts/sec)	Calculated Concentration stored curve	similar std**
Si	1.0	23.7	1.0	1.1
Cr	25.1	2259.2	24.25	26.65
Mn	1.21	84.1	0.85	1.26
Fe	52.4	2387.1	53.97	51.13
Ni	19.9	484.8	19.18	19.14
Mo	0.32	9.6	0.25	0.22

* Mo K measured at 30 kV.
** standard: type VID steel

The method is necessarily iterative. The compositions are initially assumed to be equal to the ratio of measured to pure intensity, and with these concentrations, the terms on the right of the equation are evaluated (which involve sums of concentrations and mass absorption coefficients). From these, a new concentration value for each element is obtained, and these are then used to compute new sums in the second iteration. After a few such iterations, which can be efficiently and quickly handled in a small computer, the concentration values do not change (the results have "converged") and the best estimate of concentration has been obtained. Note that in this scheme, it is only necessary to measure a single standard for each element of interest (either a series of pure elements or a single compound standard similar in composition to the intended unknowns) (Stephenson,1971).

With an energy dispersive spectrometer, since the spectrometer efficiency varies smoothly from element to element, and we may reasonably expect that the other parameters which control X-ray absorption and emission also vary gradually as functions of atomic number (we saw that this was so in dealing with electron excitation), that a curve fit through a series of measurements on pure elements, could be used to interpolate missing values (Shen,1977). Indeed, this is the case, and Figure 18.6 shows typical curves , which have a sharp discontinuity at the tube's characteristic emission line.

Although there have not yet been suitable general equations introduced (again analogous to the electron excited case) to fit well over broad atomic number ranges, largely due to the complex nature of poly-energetic excitation in which the tube spectrum contains some characteristic emission lines as well as continuum, empirical fitting and interpolation seems to work well. The curves must be determined on the same tube, or at least the same type of tube, to be used subsequently for analyzing unknowns , but this is still substantially less work than measuring intensities from many conventional standards, as will be discussed in the next chapter.

Some typical results, from published reports, are shown in Table 18.1. In some cases, the standardization was carried out using pure element standards, and in others by analyzing a single complex sample similar to the unknowns. The results are less accurate than we shall see from regression methods, but the technique can be invaluable for semiquantitative or screening analysis.

Chapter 19

Regression Models

Historically, the method used to convert measured characteristic intensities to elemental concentrations was by means of a calibration curve. By measuring intensities from a few standards (samples of known composition), and plotting them on a piece of graph paper (remember, this was long before computers became ubiquitous), a straight line relationship or at worst a smooth curve would often be obtained that could subsequently be applied to unknowns. It was understood, of course, that these curves were oversimplifications and that they could only be used when the standards were very similar to the unknowns. Any significant change in the composition of the matrix (all of the elements besides the one for which the curve was constructed) would render it useless. Nevertheless, the method often worked and had great appeal to the analyst, who could more readily comprehend the simple graph than a complex set of simultaneous equations.

It is logical, therefore, that attempts to handle the rather complex fundamental equation should have as one of their goals the possibility of representing the results as calibration curves. (And of course, we hope that the simple straight line calibration curve will emerge from all of the equations as a correct as well as useful approximation, at least in some limiting cases.)

To see how calibration curves can come about, let us start by considering a binary alloy. If the concentration of element i (the analyte) is C_i then the concentration of the other element C_j will just be $1-C_i$. The "effective wavelength" model expressed by the reduced equations we saw before may then be written as:

$$I_i/I_{pure} = C_i/(C_i + a_{ij}(1-C_i))$$

where:

$$a_{ij} = C_j(\mu(E_x)+A\mu(E_i))/(1-C_i)C_i(\mu_i(E_x)+A\mu_i(E_i))$$

again neglecting the secondary fluorescence term. The a_{ij} term is usually referred to as "α" and describes the curvature of a simple calibration curve of intensity versus concentration for the analyte element, as shown schematically in Figure 19.1. The curves must all meet at the same intensity for C=1 (the pure element), and we will call this intensity P, so that the curves have the simple form:

$$CP/I = \alpha+C(1-\alpha)$$

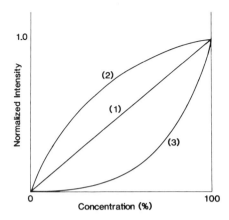

Figure 19.1 Typical elemental calibration curves of intensity vs. concentration.

This curve shows the two types of curvature which have acquired the names "absorption" and "enhancement" (although we will soon see that these are poor and misleading names, since both effects are caused by absorption -we have not yet considered flourescence at all). The same curves can be replotted in an interesting way as shown in Figure 19.2. This same transformation to a family of straight lines was shown earlier in discussing electron excitation.

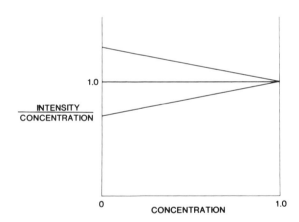

Figure 19.2 Calibration curves redrawn as straight lines.

Of course, few real analytical problems restrict themselves to binary alloys. Perhaps it is possible to get an effective α for a multicomponent system as a linear sum of the individual α s. This seems plausible since the α s themselves are simply sums of mass absorption coefficients, and we know that matrix absorption coefficients are linear sums of individual elements' coefficients. It won't be strictly true in this case because the coefficients have not all been evaluated at the same energy, and indeed the E_x energy which is the effective energy of excitation of the analyte varies somewhat

221

depending on the composition of the matrix and which other elements are present. Nevertheless, if we make the approximation, we get:

$$\alpha_{eff} = \Sigma \ (C_j \alpha_{ij})$$

where the sum is taken over all the elements in the matrix. If these approximations are acceptable, we have reduced the concentration – intensity relationship to simple calibration curves, and the α values can be computed from sums of individual two-element α s, which can in turn be evaluated from tables of mass absorption coefficients (if we are willing, as before, to estimate the effective excitation energy E_x).

Influence Coefficient Models

In a way much like the foregoing presentation, we can derive complete equations for the concentration-intensity relationship that are readily solvable by the small computers incorporated in most modern X-ray fluorescence (XRF) systems. A complete derivation of the classic "LaChance - Traill" model (LaChance, 1966), named after its first publishers, can be found in the literature, particularly in Muller (1972). For our purposes here of basic understanding, let us simply start with the fundamental equation:

$$I_i = q_i \ _{Ec}\!\int^{Eo} (C_i \mu_i (E) J(E) dE)/(C_i \mu_i^* (E_i) + C_j \mu_j^* (E_i) + ...)$$

where the dots are intended to show that all of the matrix elements must be included, and it is important to note that μ^* is not simply a mass absorption coefficient, but a sum of two, at different energies, related by a geometric factor that depends on the angles of incidence and takeoff of the X-ray beams:

$$\mu^* = \mu(E_x) + A \ \mu(E_i)$$

If we write this equation again for the pure element intensity P, there will be no matrix elements to include. The ratio of the two integrals can be simplified if we assume that the exciting radiation J(E) does not change, and that whatever the effective energy for excitation is, it does not change much from the pure element to the complex sample (this is sometimes a dangerous assumption). The integral signs will then disappear, leaving just:

$$I_i/P_i = C_i \mu_i^* (E_i)/(C_i \mu_i^* (E_i) + C_j \mu_j^* (E_i) + ...)$$

or, solving for C_i, since this is more interesting to us than the intensity we can measure:

$$C_i = (I_i/(P_i - I_i)) \ (C_j \mu_j^* (E_i)/ \mu_i^* (E_i) + ...)$$

Now, this is still a set of simultaneous equations to solve, since the concentration of all the matrix elements besides i, which are also not known, appear on the right hand side of the equation. But they are linear equations, and can be rewritten rather simply by substituting α_{ij} for the ratio μ_j^* / μ_i^*:

$$C_i = I_i/P_i \left(1 + \Sigma \left(C_j \alpha_{ij}\right)\right)$$

where the summation must be carried out over all elements except i in the matrix. This is the LaChance–Traill model, also known as a "δ – C" or concentration–correction model. The latter names come from the way in which matrix elements effect the calibration curve for the analyte. If the α terms in the summation are ignored (no interelement effects), the equation simply says that the concentration is equal to the ratio of the measured intensity to the intensity from the pure element, a simple linear calibration curve. The contribution of the α terms is to change the slope of the curve in proportion to the concentration of each matrix element.

The α factors have a reasonably clear physical interpretation, as they have been presented here. However, in practice their evaluation is difficult because the effective excitation energy is unknown. In addition, the "simple" geometric factor may be complicated by the use of a fairly broad cone of incident radiation (the consequence of placing the tube close to the sample to maximize count rate). Finally, we have ignored, so far, the effects of secondary fluorescence. The rather complex terms for secondary fluorescence do not cancel out in the ratio of integrals we started with above, and if they are of significant magnitude, will alter the result. Perhaps they can simply be swept into the α term, but how can they be evaluated?

The usual method is to measure a series of standard samples, recording intensities for all elements of importance (all those which by virtue of having a substantial concentration or a significant mass absorption coefficient, contribute to the sum of terms). Since the concentrations are known for these elements, it is possible to write a set of N (the number of standards) simultaneous equations for each element, with the unknowns being the P_i term (which it is usually not convenient to measure anyway) and the various α terms (one for each matrix element). If the number of standards is greater than the number of elements, we can solve this by straightforward least–squares regression to determine the "best fit" values of the constants. In practice, the number of standards measured should be at least twice the number of elements so that the least–squares fit can be performed with enough degrees of freedom to prevent one or a few points (which suffer the usual errors of precision due to counting statistics, sample preparation and uniformity and so on) from unduly distorting the results.

These values, also called influence coefficients, then describe how the calibration curve for the analyte element i is bent due to the presence of varying concentrations of matrix elements j. They include not only the absorption effects we have described with the equations, but the other more messy terms such as secondary fluorescence and instrumental dependence. Consequently, they cannot be easily transferred from one machine to another. The P_i term can be adjusted in proportion to changes in counting time or tube power, but not changes in tube voltage.

If a set of constants has been determined for a all of the elements in a particular class of materials, they can be stored for later application to unknowns. Then the same equations are used with the concentrations as the only unknowns. Since the concentrations appear on the right side of the equations, inside the summations, as well as on the left side of the equals sign, the equations are still not trivial to solve. Usually, an iterative approach is used. The first approximation for concentration is I/P for each element. These are used in the sums, better concentration estimates obtained, and so on until convergence is reached. The same hyperbolic or look-ahead method for speeding up convergence is used here as was described for the Z-A-F method with electron beam excitation.

Other Forms and Models

The LaChance-Traill model is widely used for routine X-ray fluorescence. However, several other models are also employed because of their particular advantages for some situations. For instance, the Lucas-Tooth (1961), or δ-I (intensity correction) model was first introduced because it was simpler to solve back in the days when computers were expensive and rare, but it has remained useful in cases where the standards are incompletely characterized. This method starts from the δ-C equations, but suggests that in the summation on the right side of the equation, the concentration of each matrix element should be roughly proportional to its intensity. We know after all that linear calibration curves are reasonable first order approximations. If this is so, then we could substitute intensity for concentration inside the summations. It will change the α constants of course, and as a reminder of this, the equation is usually written with some other symbol, as:

$$C_i = I_i/P_i \ (1 + \Sigma \ (k_{ij}I_j))$$

where again the summation is over all matrix elements (except i) which contribute significantly to the matrix absorption. These equations can be solved without iteration, since only constants or measured intensities appear on the right side. The constants P and k can be easily determined by regression or even by graphical means. For instance, rewriting the equation as:

$$C_i/I_i = 1/ \ P_i + \Sigma \ (k_{ij}/P_j \ I_j)$$

shows it to have the simple form

$$y = b + \Sigma \ (m_i \ x_i)$$

and the b (intercept) and m (slope) terms may be found in some cases by plotting the raw data on a graph and performing an eyeball or ruler fit of the "best" line.

The Lucas-Tooth model is still useful, even with more powerful arithmetic aids available, because it allows us to compute concentrations for an analyte element i without having to simultaneously determine the concentration for all of the matrix elements. In many cases, the standards that can be realistically obtained have only the elements of greatest economic interest well characterized by independent analysis. Some of the other elements, which may not even be controlled in a production process, may not be known to any degree of accuracy. A good example of this is the presence of potassium (K, Z=19) in cements. The variation of potassium from less than 1/2 to perhaps 2 percent by weight has little effect on the final product, and is not usually specified. However, since the potassium absorption edge lies just below the calcium emission energy, it strongly absorbs the calcium and some correction must be made for it to achieve the high accuracy essential for the calcium.

In cases like this, the δ-I method can be used. The intensity of the potassium can be used to scale the correction factor to be used on the calcium. Many modern programs allow some of the correction terms used to be δ-I and some δ-C, so that the term appropriate for each of the matrix elements can be selected. Of course, when intensity corrections are used, they will introduce somewhat greater errors, since the intensity is not really perfectly proportional to concentration. This will be particularly important when the concentration range of the particular matrix element is large, while if it is small, the linear approximation may be adequate (over a small range, almost any smooth curve can be adequately approximated by a straight line).

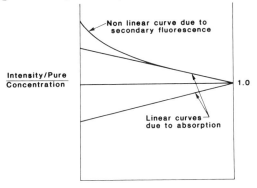

Figure 19.3 Non-linear calibration line resulting from secondary fluorescence.

Proceeding in the other direction, towards a more complex but accurate model, Rasberry and Heinrich (1974) showed that the δ-C model inadequately handles secondary fluorescence. Effectively, it treats it as a negative absorption term, that is the sign of α determines whether the presence of the matrix element decreases the intensity of the analyte (either because of a primary or secondary absorption effect) or increases it (either because of a primary absorption or a secondary fluorescence effect). All of the various absorption effects, primary and secondary, can be reasonably well lumped together, since all depend on a linear summation of mass absorption coefficients times elemental weight fractions. Secondary fluorescence behaves differently, however, as shown schematically in Figure 19.3.

The curvature of the "enhancement" due to secondary fluorescence is introduced into the regression equation as a separate term with β instead of α. Usually, a given matrix element is allowed to have only one term, either α or β.

$$C_i = I_i/P_i \ (1 + \Sigma \ (\alpha_{ij}C_j) + \Sigma \ (\beta_{ij}C_j/(1+C_j)))$$

It usually is the responsibility of the operator, in running his standards and establishing the coefficients for his "calibration curves", to select which elements to include in which summation. He may in fact omit some elements from the equation altogether if he feels they do not vary enough to effect standards and unknowns differently, or if they are minor or trace elements, or if he simply has not enough standards to be able to fit for all of the empirical constants. This process of selecting the terms to be used is called "modelling" and requires some intuitive judgement on the part of the user. He is best able to make the necessary decisions if he has some physical basis or understanding of which elements absorb or fluoresce which others. It is also useful to have the fitting program report a goodness-of-fit parameter, either the rms error, or the σ of the fit, or some other parameter (more will be said about this later on), which the user seeks to minimize by selecting the proper terms for the model.

There are still other forms of the equation. Some include higher order terms such as C^2, or cross product terms C_iC_j, and these can indeed give somewhat better fits, but the more terms that are added, the more standards will be needed. One correction that should certainly be incorporated, particularly for the energy dispersive spectrometers where background is often significant and not always easy to determine explicitly, is that for background. If the background is understood simply as the intercept of the calibration curve, then a constant term B_i might be sufficient. As we shall see, most of the background in the XRF spectrum comes from scattering of incident radiation from the X-ray tube. However, the amount of scatter depends to some extent on the matrix composition, and so it may vary from sample to sample. In addition, if there are peak overlaps, for instance of K-β on K-α (Cr on Mn in stainless steels, for instance), or L line on K line (Mo L on S K, again in steels), the measured total intensity should be corrected for this overlap. One of the methods described before when spectrum processing for electron excited energy dispersive spectra was discussed, was that of overlap factors. If that method is introduced here, the overlap terms O_{ij} can also be determined from the regression, giving as a total equation

$$C_i = I_i \ (R_i + \Sigma \ (\alpha_{ij}C_j)) + B_i + \Sigma \ (O_{ik}I_k)$$

In this form, R has been used as the rate or sensitivity of the measurement (the nominal slope of the calibration curve). The substitution of Lucas-Tooth or Rasberry-Heinrich terms for some of the α s should be self-evident.

226

Examples of Calibration Curves

If XRF measurement gave results in which the various interelement correction terms were negligible, we would call the results calibration lines, rather than curves. In fact, in many cases the degree of curvature is significant. In Figure 19.4, several examples are given for simple binary mixtures. Notice that the curvature can be in either direction, depending on the relative magnitude of the absorption coefficients for X-rays (both exciting and emitted) in the analyte element and in the matrix element.

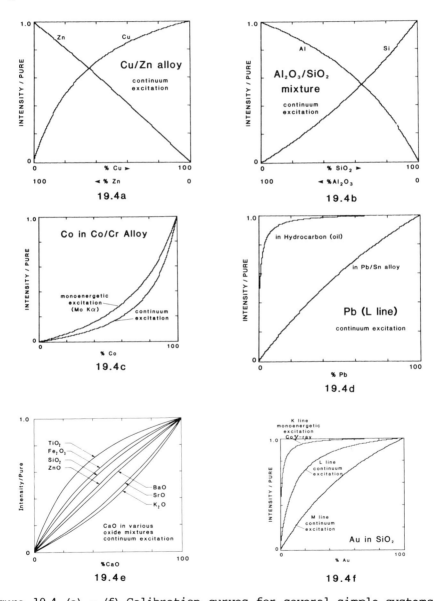

Figure 19.4 (a) − (f) Calibration curves for several simple systems.

One of the most extreme cases of a primary absorption effect is for the analysis of lead (Pb, Z=82) in gasoline. The matrix is very light and has a low absorption coefficient for both the exciting and emitted radiation. However, as the lead concentration increases, it absorbs the incident radiation stongly. The consequence is that as the lead concentration rises, the incident radiation penetrates less deeply into the sample consequently exciting fewer lead atoms. The calibration curve thus shows an initial steep region which quickly levels off, so that the measured intensity does not vary with concentration. We shall discuss depth of analysis and ways to handle these situations in another chapter.

Chapter 20

Hardware for XRF

Traditional systems for X-ray fluorescence analysis have employed wavelength-dispersive spectrometers. As described in an earlier chapter, these give high spectral resolution, and can process and count relatively high count rates (over 100,000 pulses per second) for a particular element selected by the diffracting angle of the crystal. In order to generate these high count rates, particularly for minor elements, it is necessary to use high intensities in the exciting beams, and consequently high power X-ray tubes. Close physical coupling of the tube, sample and spectrometer is also necessary to minimize the loss in intensity with the square of distance (since the radiation propagates uniformly in all directions, it spreads out over the surface of an ever expanding sphere, and the intensity per unit area falls with the square of radius, or distance).

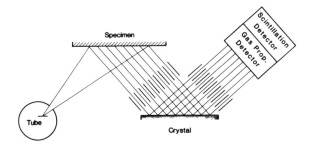

Figure 20.1 Diagram of X-ray fluorescence spectrometer.

These factors have led to the construction indicated in Figure 20.1, with only minor differences between various manufacturers. The key features become: 1) the power available from the high voltage generator (and its stability, and the ability of the tube to accept it), 2) the reproducibility of specimen position (changes of only a few tens of micrometers can cause serious misalignment of the source/crystal/detector geometry), and 3) the degree of collimation in the spectrometer (which with the choice of crystal controls its resolution), and the ability to accurately yet quickly position the spectrometer to various angles, or change crystals, to cover a series of elemental lines. The detector used will often be a gas-filled proportional detector, as described before, backed up by a scintillator and photomultiplier tube. The scintillator is simply a crystal or phosphor which emits a photon of visible light when it is struck by an X-ray. Thallium-activated sodium iodide, NaI(Tl), is almost universally used. The small light pulse is amplified by a photomultiplier tube to produce a voltage pulse that can be amplified and counted (and which has some proportionality of pulse height to energy, although with more uncertainty or spread than the gas counter). The high energy (short wavelength) X-rays that may pass through the gas counter are readily detected by the scintillator, whose efficiency falls off at lower energies. Consequently, the combination of the two detectors provides the best coverage of the range of X-rays normally analyzed.

In most modern systems, the X-ray path is evacuated to reduce absorption of low energy X-rays. The ability to detect elements below about potassium (K=19) depends on having either a vacuum path, or one in which the air is replaced by helium. The latter is sometimes preferred if the sample will not tolerate vacuum (for instance, a volatile liquid) or cannot be isolated from it in a suitable cell. These methods will be discussed further in a subsequent chapter. Given the need for an evacuated spectrometer, usually an airlock mechanism to introduce samples, or a multiple specimen carousel to permit several to be evacuated at one time (or sometimes both) are incorporated to facilitate the routine analysis of numerous samples. The mechanical ingenuity of these devices is considerable, since the specimen position must be highly reproducible. Automatic sample changers that accept dozens or even hundreds of samples from external stacks, or in one case from a conveyor belt, make it easy for continuous sample preparation to go on, with very high throughput of analytical measurements. Instruments like this allow the X-ray spectrometer to become a practical quality control device in many industries.

When high throughput is required, the time needed to scan the crystal and detector to a series of preselected angles (corresponding to the emission lines fo the elements of interest) may be unacceptably long, especially when many elements are involved. The solution to this has been to use multiple spectrometers, each with a fixed crystal and detector preset for a particular element. Simultaneous counting of many elements (as many as 20 or 30) can thus give enough counts for a high precision measurement of all the elements of interest in as little as ten or twenty seconds. With modern computers, the conversion of this data to concentration using previously determined and stored equations and constants is virtually instantaneous, and the necessary concentration information to control a process or manufacturing plant can be transmitted to those requiring it. This use of XRF systems is highly specialized, and often results in extremly costly systems that include expensive and elaborate automated sample handling and preparation equipment as well. There are even systems that bring liquid or slurry materials directly to the XRF spectrometer for in-stream analysis. We shall not consider these further, but return to the single crystal spectrometer, which sequentially analyzes the various elements, since it focuses our attention more on the X-ray portion of the system and less on the plumbing and mechanical engineering of the larger, more automated systems.

Even for the sequential spectrometers, the incorporation of stepping motors to select the crystal, position the crystal and detector, change the collimator, and so on, has become common. This allows not only a more rapid analysis, but more importantly a more reproducible one, since these functions can be placed under the control of the computer which will also handle the subsequent quantitative or qualitative calculations.

Unlike the spectrometer design used for electron excited analysis, most sequential or scanning spectrometers used for X-ray fluorescence have flat crystals which rotate about a fixed position, with gearing to move the detector by twice the angle through which the crystal moves in order to maintain the Bragg relationship. A collimator is placed in the X-ray path to ensure that only X-rays travelling in essentially parallel paths from the sample surface reach the crystal. Since the extent of the sample may be two or three centimeters, and the crystal even larger, the divergence of the beam can be quite small and still allow a reasonable intensity of X-rays to reach the crystal. The narrower the divergence angle, which is controlled by the collimator, the greater the resolution but the lower the geometrical efficiency of the spectrometer. Several collimators, each typically made from a series of thin flat sheets of metal with spacers between, and much longer than the distance between them (this design is called a Soller slit collimator after its originator), may be interchanged to trade off resolution for count rate when appropriate. The collimator essentially forces the radiation from each strip across the sample to diffract from a corresponding strip on the crystal, thus maintaining the Bragg condition.

Since with this arrangement, and with the position of the X-ray tube close to the sample as was shown before, the various portions of the sample surface are not equally excited or their X-rays analyzed, most spectrometers also rotate the sample during analysis at a low speed (a few revolutions per minute) to average the analysis over the entire area. This is only a partial compensation, however, and different sensitivities to areas at different radial differences from the center of the sample persist. For most XRF specimens, which are prepared to be homogeneous and flat surfaced over large areas, this is not an important limitation. In a few cases, such as the analysis of discrete particles on filter papers, it can produce sizeable errors.

Energy Dispersive XRF

As in the case of electron excited microanalysis, the introduction of the solid state or energy dispersive detector has resulted in some significant changes in the design of XRF systems. The most important limitation of the ED spectrometer, which arises from the nature of the amplification process and was discussed before, is its rather low count rate capability. Not only can the system handle count rates of at most a few tens of thousands of X-rays per second, compared to an order of magnitude more for the WD counting electronics (since pulses can be much narrower when it is necessary only to recognize their presence above background electronic noise, and not to precisely measure their individual pulse heights), but for the ED system, all of the X-rays generated from the sample must be processed because there is not prior rejection of X-rays from all except the single element for which the crystal diffraction angle has been set. This severely restricts the ability of ED systems to obtain useful counting statistics for adequate analytical precision for minor or trace elements, especially in the presence of major elements (even if they themselves are not of analytical interest). A good example is the analysis of low levels of sulfur and phosphorus in steels. Concentrations of less than 0.02% by weight are easily analyzed in a short time (tens of seconds) by a WD spectrometer. ED systems spend practically all of their time counting the predominant iron X-rays, which represent perhaps 99% of the sample, so that accurate results for the P and S cannot be obtained even in many hundreds of seconds. The problem of low count rate capability is compounded in this case by the poorer spectral resolution of the ED system, which makes small peaks more difficult to detect above background. We shall return to this point in discussing detection limits.

The inability to process high count rates in the ED spectrometer allows these systems to employ much lower power excitation sources. This includes the small, low power X-ray tubes mentioned previously, as well as radioactive isotopes, polarized X-ray beams, and monoenergetic X-rays produced by diffraction or secondary targets. These will be discussed later. In some cases, the low power requirements, and elimination of the need for water cooling of the tube, have allowed relatively portable XRF systems to be built using ED spectrometers.

The chief advantage of the ED spectrometer for XRF applications is its inherent simultaneity. Multispectrometer WD systems are extremely expensive and not too flexible (once set up for a particular application, they usually remain devoted to it forever). Energy dispersive spectrometers automatically collect all of the X-rays from various elements, and obtaining the counts for particular elements requires simple summation in the multichannel analyzer or computer memory. This presumes of course that enough counts will be obtained in the length of time available for the analysis to give acceptable precision in the measurement, and this in turn usually limits simultaneous ED XRF applications to those samples where the elements of interest are all major elements (eg Si, Al, Ca and Fe in cements) or where the special nature of the sample or excitation excludes any strong "major element" contributions to the total flux. Examples of the latter are trace element analysis in liquids or on filter papers, high atomic number elements in minerals (excited by high energy radioisotope sources), and so forth. Some will be discussed in a later chapter.

Qualitative Analysis

The other particular advantage of the ED spectrometer is for survey or qualitative analysis of major and minor elements (not trace elements). Qualitative analysis, the identification of the elements present in a sample, is by no means a "lesser" or easier form of analysis than quantitative determinations. The converse is in fact more often the case: it is very difficult to identify with confidence all of the elements present in a completely unknown material. Whereas with quantitative analysis, a p redetermined set of equations and constants can be routinely applied by a computer, the most valuable tool for qualitative analysis is a knowledgeable operator; automatic computer methods are far from perfected and cannot be trusted without some operator input.

Because the entire spectrum of X-rays excited from the sample is detected with the ED system, it is more difficult to overlook an element than with the WD spectrometer (where the selection of crystals and angular scanning ranges may exclude the element). True, the poorer resolution of the ED system may cause an element's lines to be overlapped by stronger lines from another, higher concentration element, but usually there will be some other peaks in the spectrum that indicate the element's presence. Also, whereas the sample preparation for WD spectrometry must produce a specimen with a flat, (and hopefully homogeneous) surface (in order for proper X-ray focus to be maintained), the ED spectrometer is not so restricted. Useful qualitative results can be obtained on very irregular and definitely heterogeneous materials, including raw mineral samples, machined or scrap metal parts, or materials moving by on conveyor belts. Sometimes, with normalization of the data, it is even possible to get rough semi-quantitative results in such cases.

The use of ED systems for XRF has increased continually since their introduction, and the greatest market has been for those applications where at least some of the need for X-ray analysis is to handle complete unknowns: samples whose qualitative composition is to be determined. This is in spite of the poorer resolution and detection limits, and is concerned mostly with the major or minor elements present in the sample. The problems associated with performing qualitative analysis by identifying the peaks present in the spectrum are identical to those discussed under electron excitation, with the added complication that whereas electrons efficiently excite a very broad range of elements because the incident electron can give up any fraction of its energy, with photon excitation a much narrower energy range, and consequently a more restricted list of elements and lines, can be fluoresced with a given source of X-rays. Consequently, the broad spectrum excitation provided by continuum radiation from low power X-ray tubes is most often used. We shall see later on that with modified excitation, other kinds of specialized analysis are also practical.

Considerable effort has been devoted to assisting the user in performing qualitative analysis, that is, identifying the peaks in the spectrum with the elements which produced them. These aids were discussed in Chapter 3, and are mentioned here as a brief review. A good start is to have available the means to display, superimposed on the spectrum, markers showing the positions and relative heights of the various emission lines. This takes about 30 lines (K, L, and M) per element to cover all of the lines with at least 1% relative height (as compared to the major line in each shell). As was described before, the use of polynomial or other functions to compute these energies allows them to be generated by a small computer program. The relative heights are more problematic, since they depend on the excitation conditions and on the matrix composition (and relative absorption at various energies). Most systems simply display the approximate relative heights for each shell, often as fixed ratios of, for instance, K-β to K-α, or sometimes with the ratio varying from element to element but using the values measured under some particular conditions of excitation, and on a particular sample, perhaps the pure element. Actual measured in tensities can vary in relative height by a factor of two or more from these "table" values, even within a single shell. Between shells, eg. L to K, the ratio can vary by orders of magnitude. Nevertheless, in the hands of a moderately skilled operator who has some independent knowledge of what elements are reasonable to expect in a particular sample, or can apply straightforward chemical reasoning, it is usually possible to identify all of the elements above trace levels, and above about Z=11 (Na) in the periodic table, using ED spectrometry, in a few minutes. A WD system could perform the same job, but the sequential scanning of all possible elements would take many times longer (perhaps 1/4 to 1/2 hour), and the resulting strip-chart recording would have to be manually interpreted. (Most modern systems would automatically use pulse height discrimination to reject high order diffracted lines, which would otherwise further clutter the spectrum.)

The strategy for qualitative analysis which seems to work in most general cases is to start with the largest peak in the spectrum. This almost certainly must be an α line (there are a few instances in which the L-β or M-γ may be larger than the α, but they are unusual). Identifying this line with possible elements is straightforward, since the table of line energies is not too large, and most systems allow it to be searched by either atomic number or energy (wavelength). In many cases, there will be only a single K-line match, and at most one or two L or M matches, even taking into account the greatest possible errors in spectrometer calibration. Deciding between the few possibilities is then based on the pattern of other lines in the spectrum: a K-α line is accompanied by only a K-β line, generally only about one-sixth as large. This varies somewhat with atomic number, being smaller for lower atomic numbers where the M shell of electrons (whose transitions to the K vacancy generate the K-β) is not filled. It may also, at higher energies, become possible to partially resolve with the ED spectromer the K-α-1/α-2 doublet, or the several K-β lines. But in any case the appearance of a set of K-lines looks substantially different than the family of L lines, in which the L-β1 and 2 (at least), and usually the L-γ-1 and L-η, are quite distinct, spread over a considerable energy range, and easily recognized. The M lines may sometimes be confused with the L, becasue the resolution of the ED system at low energies is insufficient to clearly distinguish them. But in these cases the presence of the higher energy L lines will serve to identify the M-line element. Figure 10.2 illustrates the appearance of lines from the various shells, at the same energy in the spectrum.

In summary, then, the element with the largest peak in the spectrum can usually be identified. Once it has been, all of the peaks which correspond to that element can be marked and disregarded. The process is then repeated with the largest remaining line, and so on. Eventually, you get down to the point where the only peaks left are trace elements for which only the largest peak can be seen above background, and then the identification becomes more troublesome in the abscence of minor lines to confirm the element.

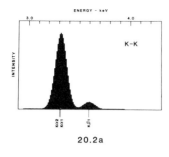

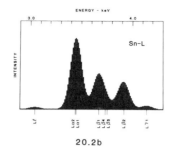

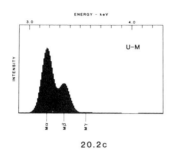

Figure 20.2 Spectra from (a) potassium K, (b) tin L and (c) uranium M
lines all lie at about the same energy.

The principal difficulty arises when overlaps of lines from one
major element hide one or more lines from another, subsequently
identified element. It may then not be possible to distinguish between
several elements positively. It may even happen that the hidden line
is the α line, and the only lines left are the smaller lines. This will
not be apparent if the element is a minor one, and the change in the
relative size of the major element line due to the included overlapped
peak, is not large enough to warrant attention. Situations like this
place the burden back on the instrument operator, to apply independent
knowledge and judgment based on the chemical or physical nature of the
sample to distinguish between the elements.

It is also possible to miss certain elements altogether if the
operator does not "expect" them, because all of the peaks are hidden
by more major elements. A common example is that of manganese in
stainless steels. The only manganese peaks detected are the K-α at 5.9
and K-β at 6.5 keV. With the resolution of a typical ED system, the K-α
will be hidden by the larger K-β peak at 5.95 keV. from the chromium,
whose concentration is ten to twenty times greater than the manganese,
and the manganese K-β will be hidden by the larger K-α peak at 6.4 keV
from the iron, which is the major element in the sample. Without
detailed peak fitting to recognize the distortion in the shape of
these peaks due to the presence of the hidden manganese peaks, or the
knowledge of the operator to expect manganese in samples that contain
large amounts of iron and chromium, the manganese would go undetected
even at concentrations of more than 1%. This was illustrated in
Chapter 3.

The spectrum appearance and artefacts are similar in energy-dispersive XRF analysis to that from electron excitation, with a few exceptions. The much lower background intensity with photon excitation permits smaller peaks to be seen, including both peaks from lower concentration elements, minor lines from elements with large peaks, and artefacts such as escape peaks. The spectrum may also contain scattered peaks from the excitation source (characteristic lines from the tube target -both the principle element and and contaminant elements, or monoenergetic lines from a radioisotope, for example). Finally, diffraction peaks may appear if the sample material is crystalline and continuum excitation is employed. The band of energies (wavelengths) of incoming X-rays may include some for which the Bragg equation

$$\lambda n = 2 d \sin (\theta)$$

is satisfied, and these will be diffracted to the detector, measured and recorded as peaks in the spectrum. Since a range of θ angles is covered by the geometric arrangement of source, sample and detector, these peaks are generally somewhat broader than the normal emission peaks (this may not be true if the sample is highly oriented or a single crystal material). It is also generally possible to recognize diffraction peaks by slightly changing the specimen position, for instance by raising it in its holder. This changes the angle from source to detector so that a different wavelength (energy) is diffracted, and so the peak(s) shift in the spectrum, whereas emission peaks do not. Nevertheless, the presence of the diffraction peaks can confuse spectrum interpretation and processing, and may hide other peaks of interest.

There have been occasional attempts to use this phenomenon as a tool to study structure of samples by "simultaneous" X-ray diffraction (Giessen, 1968). The resolution of the diffraction spectrum, even with slits to restrict the range of θ angles to a small fraction of a degree (much the same as in a conventional X-ray diffraction spectrometer), is much too poor to permit general purpose identification of unknowns. Several specific applications to quantify the amount of a few known phases (for instance, austenitic and ferritic phases in stainless steels) or to monitor phase changes during the application of pressure or temperature variations, have been reported successful.

Chapter 21

Sample Effects

Most real samples have surfaces that are not perfectly flat and homogeneous. In the WD spectrometer, roughness or out-of-flat conditions will reduce intensity because the generated X-rays are not properly diffracted to the detector. A change in the sample surface position of a few tens of microns can produce a 10% drop in intensity. Similarly, scratches on the surface, for instance from grinding or polishing, can reduce the intensity. Figure 21.1 shows the nature of the effect, in which the reduction in characteristic intensity for pure, massive samples is plotted for various elements as a function of groove size. The magnitude of the mass absorption coefficient of the matrix for the particular energy of the radiation controls the magnitude of the effect. Aligning the sample so that the grooves run parallel to the X-ray optic path, as shown in the second sketch, will reduce the effect to negligible. This is another reason that most WD XRF systems employ sample rotation.

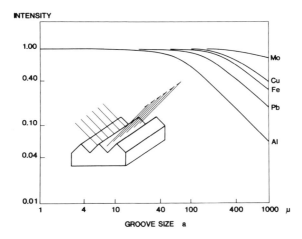

Figure 21.1 Effect of transverse grooves in sample surface on intensity (Muller, 1972).

Many X-ray fluorescence samples are prepared by grinding the material to be analyzed to a fine powder, which is then compacted to form a solid disk. The various components in the raw material may not grind uniformly, if their hardnesses or initial sizes are different, giving rise to a final sample made up of a mixture of particle sizes, where each component with a different composition also has a different particle size. The results of this heterogeneity on measured intensities can be quite different, depending on the relative composition of the coarse and fine fractions.

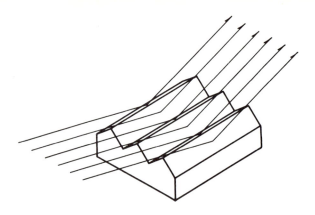

Figure 21.2 Longitudinal grooves in surface (parallel to plane of X-ray optics) do not reduce intensity.

First, consider the case of homogeneous powders. The standard and unknowns may still have different particle size distributions, which will pack differently to produce different intensities from the same composition. Generally the finer the powder, the greater the packed density and the higher the intensity. It is the aim of sample preparation (grinding, pressing or solution) to make all samples equally dense. Figure 21.3 shows the intensities of silicon and calcium as function of grinding time for a cement raw mix. For routine operation, a time would be selected that lay on the plateau of intensity, where the powder was so fine that the depth of analysis covered many grains and the individual particle size was unimportant. Similarly, the effect of increasing pressure used in forming the pellet or disk to be analyzed effects the measured intensity. Figure 21.4 shows the effect of pressure on powders of various particle sizes, for a particular element. Again, the goal of sample preparation is consistency, and this is best obtained with very fine powders and high pressures.

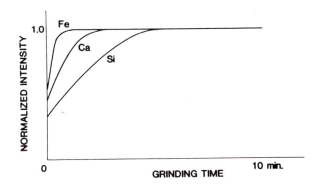

Figure 21.3 Effect of grinding time on measured intensities in cement raw mix (Bertin, 1970).

238

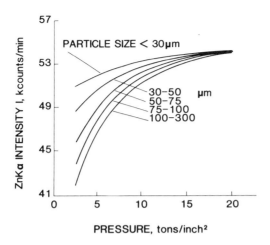

Figure 21.4 Effect of pelletizing pressure on intensities for different particle sizes (Bertin, 1970).

The figures illustrate what happens with finite particle size. The analyzed region extends to a finite depth, controlled by the absorption of incident exciting radiation and emitted characteristic X-rays in the particular matrix. As the particles become finer and finer, the analyzed volume increases, as a fraction of the volume of a completely dense solid.

For heterogeneous powders, which result from most real materials since the different phases or components have different composition and physical properties, and respond differently to grinding or other sample preparation steps, the situation is more complex. If the particles are very coarse, much larger than the depth of analysis, the X-ray intensity that is measured will represent the area fraction of each phase on the exposed surface. If the various species are roughly the same size, this will give a straight line relationship of intensity versus volume fraction. If one phase is much finer than the others, it will be preferentially packed in interstices and exposed on the surface, giving much too high an intensity (Figure 21.5). The measured intensities in this case would give a roughly constant signal from the elements in the coarse particles, regardless of the amount, since their size determines the amount of the phase exposed. The intensity from elements in the small particles would be proportional to the volume they represent.

Figure 21.5 Schematic diagram of the effect of finite particle size on X-ray intensity.

For the more general case, the situation is more complex. Figure 21.6 shows the effect of particle size on the shape of the calibration curve for a few cases. Only after extended grinding to obtain fine particles do the "normal" absorption or enhancement type curves appear. Mathematical models to correct for various particle size distributions have not met with general success. Monte-Carlo techniques can in principle predict the intensities from complex composition and size distributions, but this is impractical for real analytical situations.

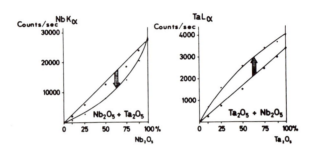

Figure 21.6 Effect of particle size on the shape of calibration curves (Muller, 1972).

Depth of Analysis

The emphasis in discussing particle size was on the need for sample preparation techniques to reduce the size of particles (or depth of surface irregularities, or any other measure of heterogeneity in the sample) to well below the depth of analysis for the analyte element, in the particular matrix. For the geometrical arrangement shown in the figure below, the depth of the layer from which 99.9% if the analyte line intensity comes can be estimated (in centimeters) as Bertin, 1970):

$$6.91/(\mu(E_x)/\sin \phi + \mu(E_i)/\sin \theta) \rho$$

where μ is the matrix mass absorption coefficient (the sum of the individual coefficients times their respective weight fractions), and E_x and E_i are as usual the effective excitation energy (unknown, but for continuum excitation generally just above the critical excitation energy E_c) and the emission line energy of the analyte, respectively. Using this expression, it is possible to estimate the major element composition of the samples to be analyzed, calculate the depth of analysis for each element, and use the least of these as a criterion for sample preparation. If the sample heterogeneity can be reduced to significantly less than this distance, say to one tenth the depth, then for practical purposes, it can be assumed homogeneous. Figures 21.7-21.10 show the depth of analysis of elements in several different matrices.

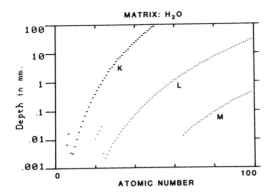

Figure 21.7 Depth of analysis for various elements in aqueous solution.

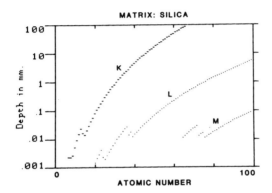

Figure 21.8 Depth of analysis for various elements in silica.

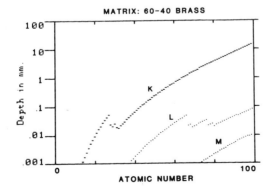

Figure 21.9 Depth of analysis for various elements in brass.

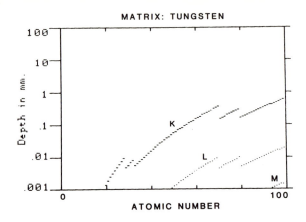

Figure 21.10 Depth of analysis for various elements in tungsten.

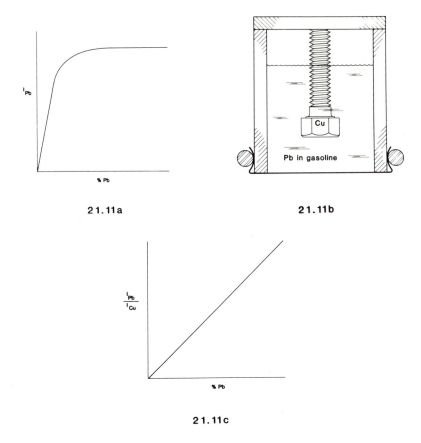

21.11a

21.11b

21.11c

Figure 21.11 (a)-(c) Introduction of internal reference into specimen holder for gasoline analysis.

242

Another subject related to the depth of analysis, is the problem introduced before in discussion of the primary absorption effect of a heavy element (Pb) in a light matrix (gasoline). As the concentration increases, the depth of analysis is reduced because the lead atoms largely control the absorption of the incident, exciting radiation. Consequently, the measured emitted intensity for lead becomes nearly constant, and not proportional to the lead concentration. One solution, which can be well adapted to this and other analysis where the matrix is a liquid or loose (uncompacted) powder is shown in the sketch in Figure 21.11. The introduction of the metal screw in the top of the liquid cell (it can be copper, iron, or any other element with a reasonably high characteristic energy) produces X-rays whose intensity gives us the needed information. Instead of monitoring the intensity of the lead, we can use the ratio of the lead to the copper (for example). This ratio will continue to increase with lead content, because as the lead concentration increases, the depth of excitation will drop, and the generation of the copper X-rays will decrease. Also, the absorption of those X-rays in the matrix will increase, so that the total intensity will drop. Consequently, either this intensity by itself, or a ratio of the lead to the copper, can be used to construct a working calibration curve for the lead content in gasoline. From the expression already given for the depth of excitation in a given matrix, we can estimate that the distance from the surface to the screw or pin should be about a centimeter, depending on the range of lead we wish to analyze, and the excitation spectrum.

Analysis of layered structures

Other kinds of samples of considerable practical interest are layers of one material (paint, plated metal, oxide coating, etc.) on a substrate. Sometimes we wish to analyze the concentration, sometimes the thickness, and sometimes both. If the thickness is in the range where some penetration of exciting radiation to the substrate takes place, and the emitted characteristic X-rays can be detected (from some element not also present in the coating), then we can use the X-ray fluorescence method to determine the desired information. Sometimes it is possible (or necessary, if the coating is an organic material) only to monitor the substrate intensity and from its reduction, determine the coating thickness. In general, we will try to use intensities from both the coating and substrate elements.

For the substrate, which we are assuming in this discussion to be infinitely thick, the measured intensity is controlled by the absorption of both primary and secondary radiation. Relative to the intensity measured from an uncoated surface, the intensity as a function of mass thickness pt (density times thickness) is:

$$I/I_O = e^{-(\mu(E_x)/\sin(\phi)+\mu(E_i)/\sin(\theta))\,\rho t}$$

where all of the terms have their usual meanings. For the elements present in the coating, the intensity relative to that from an infinitely thick (ie. thicker than the depth of analysis) layer is:

$$I/I_{inf} = (1/\mu^*) \, {}_0\!\int^t e^{-z\mu^*}dz$$

where

$$\mu^* = \mu_{mtx}(E_x) + A\,\mu_{mtx}(E_i)$$

and $A = \sin \phi / \sin \theta$

as usual. These two expressions give curves of intensity versus coating thickness which are generally close to straight lines on semi-log paper (since absorption is an exponential function of distance), as indicated in Figure 21.12. The ratio of coating element intensity to substrate element intensity is similarly often plotted on a logarithmic scale. Regardless of the way in which the data are represented, if enough standard samples can be obtained to independently vary the coating composition and thickness, a family of curves can be established to determine either or both. The example shown in Figure 21.13 is for an unsupported iron-nickel foil, for which both the thickness and concentration can be determined uniquely from the ratio and sum of the two elements' intensities.

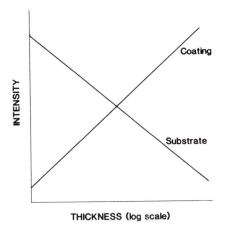

Figure 21.12 Variation of intensity from coating and substrate, with thickness.

Filter Paper Samples

A rather special variant of coating analysis is the measurement of elements on filter papers. These are typically deposits from air or liquid (especially water) collected on a low mass, low atomic number substrate (most often simply paper, but sometimes other kinds of membranes or organic films). Both simple mechanical filtration (Campbell, 1966), as from high volume air samplers used for pollution measurement, and ion exchange or other chemical precipitation techniques (van Niekerk, 1960) can be used to accumulate the sample on the paper. Both of these methods have been especially used with ED spectrometry, because of the ability to detect simultaneously a very broad range of elements (and because the count rates are usually inherently rather low from these kinds of samples). The latter technique has been particularly developed for trace analysis of metals in water. Passing the water several times through a paper loaded with an ion exchange resin (with some attention to maintaining the proper pH) collects essentially all, or at least a constant high fraction of the ions in the water. The analysis of the filter paper then gives a linear calibration curve of intensity versus the amount of the element on the paper, which in turn is linearly related to the concentration in the liquid. With this kind of chemical concentration, extremely low trace amounts of the elements in the original solution can be analyzed, often in the part-per-billion range.

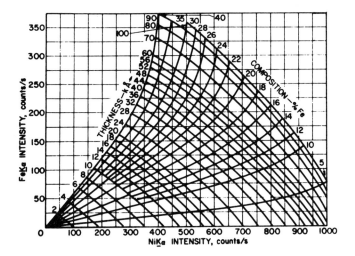

Figure 21.13 Calibration curves for concentration and thickness of iron-nickel foils (Bertin, 1970).

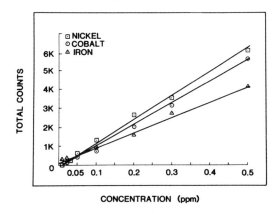

Figure 21.14 Calibration curves for trace elements in water.

Most analysis of filter papers, either using chemical or simple physical means to collect the sample, do offer this ability to analyze low original concentration levels. Also, because the amount of material on the filter paper is usually a low mass thin layer, it is thin compared to the depth of analysis which would be calculated using our earlier expression. Consequently, the effects of matrix absorption on the incoming or outgoing radiation are negligible, and without absorption there can hardly be any secondary fluorescence. Thus the calibration curves for elements on filter papers are usually linear, at least up to moderate loading levels. Figure 21.15 shows, for different atomic number elements (Fe & Si), the linearity of the calibration curve up to nearly 1000 micrograms per square centimeter, beyond which the effects of absorption begin to cause some curvature.

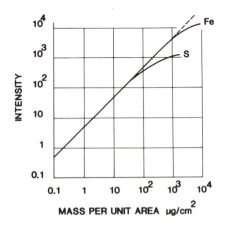

Figure 21.15 Calibration curve for trace Fe & Si on filter paper, showing extent of linear range.

The consequence of this simple linear behavior is that samples are easy to prepare. Elements can be deposited in known amounts (mass per unit of area) but in very different physical form than expected in the unknowns. This could take the form of chemical solutions evaporated to leave a residue, known weights of stoichiometric compounds deposited as fine powders, or even thin metallic foils. All have been used, and the choice must be based primarily on the convenience with which the particular elements of interest can be prepared in one or another way, and the precision with which the standard can be characterized.

It has not been the purpose of this brief chapter to catalog the various sample preparation techniques, which are well covered in the existing literature. They are identical for ED and WD XRF, since the physical processes in the sample are the same, and are intended to produce samples for which the quantitative models can adequately describe those physical processes. Rather, it is hoped that the concepts presented will stimulate the analyst to consider the ways he can measure X-ray intensities from various depths in complex structures, or estimate the effect of inhomogeneities, so as to be better able to choose appropriate mathematical models or preparation techniques.

Chapter 22

Scattering, Background and
Trace Element Analysis

Most of the background in the photon-excited X-ray fluorescence spectrum is caused by scattering of X-rays originating in the X-ray tube, from the sample to the detector. A very minor amount of the measured backround is our old friend Bresstrahlung, generated in the sample by the deceleration of photoelectrons knocked out of atoms in the sample when they are excited. But since most of the instruments actually in use have an X-ray tube producing the exciting radiation, and since its output includes a substantial amount (often it is the dominant fraction of the total radiation) of continuum, it is scattering that is responsible for most of the observed spectrum background.

Scattering of X-rays is one of the phenomena that were mentioned as being responsible for the X-ray mass absorption coefficient. While at the energies we usually deal with in X-ray analysis, the major component of absorption is the photoelectric effect (absorption of the X-ray by ionization of an atom), scattering can also occur. Actually, there are two kinds of scattering, which add together. One is coherent (or elastic, or Rayleigh) scatter, and the other incoherent (or inelastic, or Compton) scatter.

In coherent scatter, the incident X-rays force the bound electrons in an atom of the sample to oscillate at the same frequency as the photon, and the vibrating electrons then re-emit the same X-ray photon in another direction. There is no loss of energy. Bragg diffraction is a special case of coherent scattering when the re-radiated waves from many atoms in the sample are in phase (add constructively) in a particular direction. Coherent scatter will cause some of the X-rays from the tube (or other excitation source) to be redirected to the detector. The efficiency of this process (the elastic scattering cross section) depends on X-ray energy, so that the shape of the exciting spectrum will not be reproduced exactly in the measured spectrum, but the individual X-rays are not changed in wavelength or energy, so if there is a strong emission line in the exciting spectrum, some of those X-rays will show up as a peak in the measured spectrum.

Incoherent scatter is at first glance not different than the coherent scattering just described. The incoming photon causes an electron belonging to an atom in the sample to oscillate at its frequency, and this then produces a re-radiated photon in another direction. But if the electron lies in an outer, loosely bound orbit around the atom, it can recoil in the process. The energy absorbed by the electron knocks it away from the atom, and removes energy from the

247

photon, so that the re-radiated (scattered) X-ray has less energy, or a longer wavelength. Figure 22.1 shows this process schematically.

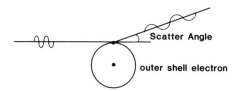

Figure 22.1 Incoherent scattering redirects X-ray photons with loss of energy.

It is clear that energy is conserved in this process, since the electron's kinetic energy accounts for the loss in energy of the photon. But momentum must also be conserved, and this means that depending on the angle of scatter ϕ , at which we observe the scattering, the direction and energy of the electron, and consequently the amount of energy loss of the scattered photon, are forced to take a particular value. In terms of the wavelength λ , the change (increase) in wavelength is given by:

$$\delta\lambda = (h/mc) \ (1-\cos \phi)$$

where h is Planck's constant, m is the electron's mass, c is the velocity of light, and substitution of numeric values reduces this to:

$$0.0243 \ (1-\cos \phi)$$

for the change in wavelength in angstroms. This change in wavelength or energy is independent of the original energy of the photons, and does not depend on the nature or composition of the target atoms. The angle ϕ is determined by the geometry of the X-ray spectrometer, and is usually close to 90 degrees (the angle between the incoming beam of photons which excite the sample and the path of emitted X-rays toward the spectrometer). At this angle, the total amount of scattering is minimized. The amount of scattering (the cross section for incoherent scatter) varies with X-ray energy, and also with the composition of the target (the number and bonding strength of the outer electrons).

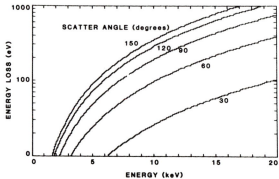

Figure 22.2 Energy loss by incoherent scattering at different angles and energies.

For a 90 degree scatter angle, the change in wavelength of incoherently scattered photons is 0.024 angstroms. In an energy spectrum, the shift in energy varies with energy as shown in Figure 22.2. Different shifts are observed for varying source/sample/detector angles, and when a range of such angles are present, scattering will occur at different angles with a concurrent range of energy loss amounts.

The amount of total scattering due to incoherent, as opposed to coherent scatter, increases when the target (the sample) contains a large proportion of atoms with loosely bound outer electrons, or in other words as the atomic number of the atoms decreases. Consequently, low-Z samples produce the greatest amount of incoherent scatter. For a high atomic number material such as lead, virtually all of the scatter is coherent. For a low atomic number material such a graphite, the ratio of incoherent to coherent is about 6:1. The fraction of incoherent scattering also rises with energy, particularly when the X-ray energy is substantially greater than the binding energy for the outer shell electrons.

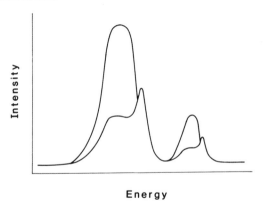

Figure 22.3 Scattered peaks from a Mo target X-ray tube, showing effect of low and high Z materials. The higher incoherent scatter curve is produced by using water as a sample, and the lower one using brass.

Since most X-ray tubes will produce a substantial characteristic emission line from whatever their target element happens to be, we should expect to see this energy appear as a peak in the measured spectrum, due to scattering. The spectra shown schematically in Figure 22.3 illustrate the use of several different target materials to scatter the radiation from a molybdenum target X-ray tube. We are not seeing the emission lines from the elements in the sample, which lie at different energies, but simply the scattered molybdenum K-α and K-β lines. Note that the relative magnitude of the incoherent (Compton) peak increases for the low atomic number samples. Also, the breadth of the coherent peak is about what we would expect for the resolution of the energy dispersive spectrometer at this energy, whereas the breadth of the incoherent peak is much greater. This is because the sample and the detector are of significant size, and there is a range of possible angles through which scattering can occur to redirect photons from the source to the detector. Each different angle is accompanied by a slightly different loss of energy, and so the total breadth of the peak is increased and its shape may also be distorted.

Because of the presence of peaks due to the scattered characteristic lines of the X-ray tube, the presence of trace amounts of those same elements may be difficult to ascertain. Furthermore, the Compton scatter peak from the characteristic tube line may also interfere with another element, often the next lower atomic number. One criterion for selection of the tube target material will then be to avoid this problem, and selecting an element whose presence is not of interest (or conversely, one whose concentration is so high that the small amount of scattered background radiation will be insignificant as compared to the emitted characteristic line from the sample).

For most elements in the sample, the background under the measured spectral peak will be due to scattered continuum radiation from the X-ray tube. The presence of the background, and the counting statistics associated with it,makes it difficult to detect with assurance the presence of very small peaks from elements in the sample. The detection limit can be variously defined, but amounts in general to a concentration level whose presence can be expected to give a peak which will rise above the local background by an amount great enough to be discerned with a certain statistical confidence factor.

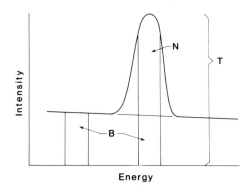

Figure 22.4 Diagram of peak and background intensities discussed in the text.

As shown in the sketch of Figure 22.4, the background intensity must usually be estimated from the spectrum on one or both sides of the peak. For a total intensity T for peak and underlying background, a background intensity of B, and a net intensity N=T−B, measured in t seconds on a sample of concentration C, the detection limit may be defined as:

$$n \, C \sqrt{(B)} \, / \, N$$

where n is a factor of confidence, often taken as 2 (2 standard deviations, or 95%) or 2.33 (99%). This expression allows us to estimate the concentration level whose peak we would probably (hence the "confidence level") be able to detect if it were there, and is hence the concentration that corresponds to the upper limit of a "not detected" report. The lowest level we can be sure of detecting is typically 2–3 times higher. It is somewhat more convenient to rewrite this expression in terms of the background count rate I_b, the analysis time t, and a factor called the "sensitivity" of the measurement, m which has units of net counts per second per percent. Then, for a "two-σ" (95% probability) detection limit, we would have:

$$2/m \ (I_b/t)^{1/2}$$

This expression is appropriate for energy dispersive or multiple crystal diffractive systems, where the background and peak intensities can be measured simultaneously. For a sequential or scanning diffractive wavelength dispersive system, since an equal amount of time t would have to be spent measuring the background, the detection limit would be worse (higher) by the square root of two. (The subject of optimal counting time strategies for counting the peak and the background has arisen many times in the literature of WD systems. We shall ignore it here, because at the detection limit, the two counting times should be the same in any case, and furthermore this does not concern us with ED systems.)

From this expression, we can observe several important relationships. First, the detection limit will improve in proportion to the square root of counting time. If we count 4 times as long, the detection limit drops only by a factor of two, or a factor of 100 in counting time gives only a factor of ten in detection limit. This often leads to cases of very diminishing returns in trying to reach low detection levels by simply extending the analysis time. The role of the background level and the factor m is also interesting. Reducing the background will directly reduce the detection limit, and indeed most work toward improved detection limits is concerned with reducing the intensity of the background under the peak(s) of interest. The value m is just the slope of the calibration curve of intensity versus concentration. Near the detection limit, this is essentially a straight line. A high slope (high sensitivity) improves the detection limit, and this generally means that we want to design the system or tailor the excitation to produce the highest possible count rate and the most efficient excitation possible for the elements of interest, to achieve good detection limits.

In X-ray fluorescence using X-ray tubes for excitation, without resorting to the tricks of modifying the exciting radiation with filters, secondary targets, polarization and so on which will be discussed in another chapter, the background is usually fairly low (especially as compared to electron excitation with its abundant Bremsstrahlung production in the sample). Detection limits are generally in the range of a few parts per million (by weight), as compared to perhaps a hundred times that for electron excitation. Of course, since a much larger sample volume is being analyzed, this still represents a larger total mass of the element than can be detected when excited selectively by the electron beam.

Quantification at Low Levels

At the trace level, or indeed for some elements and matrices for concentration levels up to much higher amounts, the presence of the analyte has no appreciable effect on the matrix. This is equivalent to saying that the matrix mass absorption coefficients for the element's emitted radiation, and for the incoming radiation that excites it, which are formed from sums of each element's weight fraction times its individual absorption coefficient, does not change when the element is included in the summation. It is this insensitiviy to the element's presence that produces a straight line calibration curve.

But it is important to remember that the calibration curve we see as a straight line at low concentration (which may be the only range we care about for a particular element and type of sample) is only a part of the complete curve, which will in general have the same shape as we saw previously. If we treat the entire matrix as the other component in a psuedo-binary alloy, the curve relating intensity I_i for the analyte to its concentration C_i is given by:

$$I_i/P_i = C_i \, / \, (C_i + r_{ij}(1-C_i))$$

where r_{ij} is the ratio of. μ^*_j/μ^*_i

and $\mu^* = \mu(E_x) + A_{geom}\,\mu(E_i)$

using the same notation as previously.

As concentration approaches zero, the slope of this curve is simply $1/r_{ij}$, and this corresponds to the slope m of the calibration curve used above to estimate detection limits. A variety of calibration curves may apply to different elements in the same matrix, or the same element in different matrices, as indicated in Figures 22.5 and 22.6.

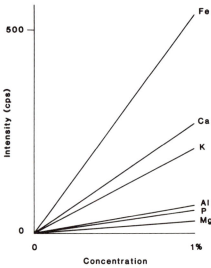

Figure 22.5 Calibration curves for trace levels of different elements in the same matrix (Muller,1972).

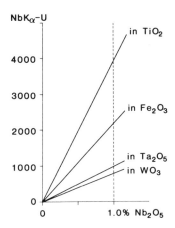

Figure 22.6 Calibration curves for trace levels of the same element in various matrices (Muller,1972).

There will also be an effect of excitation, since the slope of the calibration curve will be different for different tube targets. This can be visualized as a change in μ^* due to a shift in E_x when the tube spectrum has a strong characteristic line positioned to strongly excite the element. For instance, Figure 22.7 shows calibration curves for the Ta L-α line in a mixture of Ta_2O_5 and Nb_2O_5, using W and Mo target X-ray tubes. The tungsten tube can excite the Ta only with continuum , and hence E_x is quite close to E_c. The molybdenum tube has a strong characteristic K-α line which can excite the tantalum, and this raises E_x farther above E_c where μ is lower. The change in slope of the curve approaching zero concentration is quite different, and reflects an improvement of nearly a factor of 2 in detection limit for the Mo tube.

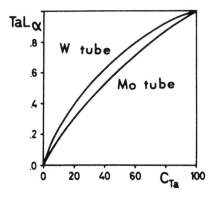

Figure 22.7 Effect of excitation on calibration curve shape (Muller,1972).

Since the portion of the calibration curve needed for quantification at low levels is treatable as a straight line, all we need to do is determine its slope. This can be found in several ways:

1) pure calculation based on the absorption coefficients, 2) measurement from standards with a similar matrix, 3) addition, or spiking techniques which add more of the element to the sample, and 4) calculation relative to another matrix for which measurements have been made. The first of these is rarely used , as it requires a complete knowledge of the matrix composition and a good estimate of E_x, the effective excitation energy. The second method is straightforward, but like any calibration curve technique may be impractical if appropriate standards are not conveniently available.

The third method is useful with liquids or some loose powders. The unknown is first measured, to obtain a net intensity I_u corresponding to its (unknown) concentration C_u. Then a known amount of the element is added to the sample, which is again mixed for homogeneity and again analyzed. The intensity this time (I_1 is greater. If the operator's guesswork is good, the concentration has been about doubled and so has the intensity. More additions and measurements can be made, to allow good fitting of the calibration line shown in Figure 22.8. From the intercept of this line, the concentration of the original sample can be determined straightforwardly.

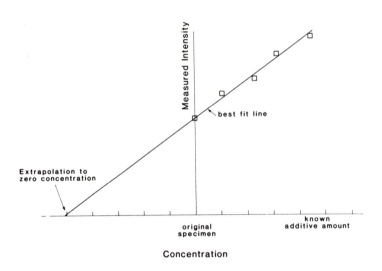

Figure 22.8 Linear calibration curve produced by the linear addition technique.

Calculation from another matrix is a powerful method when the matrix composition is known, but no similar standards are available. As shown in Figure 22.6, the slope of the calibration curve for a given element is different in different matrices. If it can be assumed that E_x is the same for both matrices and there are no other elemental absorption edges between E_c and E_x, then the ratio of the measured intensity for the element in the first matrix to that in the second is just the ratio of the integrals from the fundamental equation introduced earlier, which incorporates mass absorption coefficients evaluated at E_i and E_x. But neglecting absorption edges, the variation of mass absorption coefficient with energy is as we saw earlier:

$$\mu(E) = \text{const. } z_i{}^m / E^n$$

where m and n are "constants" (n is approximately 2.8). This means that the total matrix absorption coefficient is just proportional to the sum (for all the elements) of

$$c_j z_j{}^m / E^n$$

and consequently the ratio of $\mu*(\text{matrix 1})/\mu*(\text{matrix 2})$ becomes a constant. Then the ratio of intensities in just proportional to the ratio of mass absorption coefficients, which can conveniently be evaluated at the emission energy E_i. Hence, to a first approximation, the slope of the calibration curve will just be inversely proportional to the matrix mass absorption coefficient for the analyte element. Figure 22.9 shows this relationship for two elements (200 ppm of zinc, and 1% of niobium oxide) in several matrices. The intensity (which is just proportional to the slope of the calibration curve) is plotted against the reciprocal of the matrix mass absorption coefficient, giving a straight line. This would allow us to determine the calibration curve for either element in any matrix, if we had measured it in one, and knew the composition (in order to get the mass absorption coefficient) for the other.

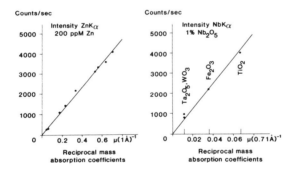

Figure 22.9 Slope of calibration curve plotted against reciprocal of the matrix mass absorption coefficient.

It is also sometimes possible to estimate the matrix mass absorption coefficient at a particular element's energy without knowing the composition of the matrix. One method is to estimate it from the height of the scattered background. We saw before that the ratio of incoherent to coherent scatter varies with atomic number, because of the change in the binding energy and number of outer shell electrons. This can also be plotted as a function of total matrix mass absorption coefficients, as shown in Figure 22.10, since the total absorption coefficient also varies with atomic number (recall the Z dependence just discussed). This suggests that if we can determine the ratio of coherent to incoherent scatter (for instance from the scattered and recorded characteristic lines from the X-ray tube target element), it should be possible to estimate the mass absorption coefficient of the matrix for any element relative to any other, based on the relative (scattered) height of the spectrum background (Van Dyck,1980).

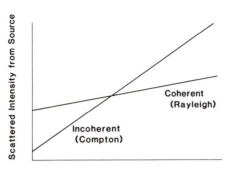

Figure 22.10 Atomic number dependence of the ratio of incoherent to coherent scatter.

This technique has indeed been developed by several authors, and is particularly well applied to the analysis of low levels of elements in mineral specimens. The technique is able to produce 5 to 10% relative accuracy for the elements, based on analysis of several of the National Bureau of Standards trace element samples, as shown in Table 22.1.

This technique requires a fairly elaborate computer program to determine the needed absorption coefficients, and is most practically applied using energy dispersive techniques in which the entire spectrum background and the scattered tube characteristic lines can be recorded. Just as in the case of electron excitation, we were able to use the spectrum background to gain information about the sample matrix, so too with X-ray excitation we can sometimes extract useful information from what is normally a nuisance component of the total spectrum.

Another rather specialized method to determine the matrix mass absorption coefficient for a particular sample and specific elements is to measure it directly, using the sample. This is done as shown in the schematic diagram of Figure 22.11. First the spectrum from a standard specimen, containing known amounts of the (minor) elements of interest, is measured. Then a thin section of the sample, of known mass per unit area (the product of density and thickness) is placed between the bulk sample and the detector and the intensities measured again. From the reduction of the intensities for the elements of interest, and the known sample mass thickness, the absorption coefficient can be determined from Beer's Law $I/I_0 = \exp(-\mu\rho t)$.

This procedure is repeated for the unknown sample. In all, four sets of intensities are measured, which yield the intensity for each element in the bulk standard and in the bulk unknown, and the matrix

Table 22.1 Trace concentrations determined by using scattered intensity to estimate matrix absorption coefficient (Giauque, 1979).

(concentrations in $\mu g/g \pm 2\sigma$)

Element	NBS SRM 1571 Orchard Leaves		NBS SRM 1577 Bovine Liver	
	XRF	NBS	XRF	NBS
Ti	18.0±8.5	–	<11	–
V	<8	–	<6	–
Cr	<5	2.6±0.3	<4	–
Mn	86.5±4.9	91±4	9.4±1.1	10.3±1.0
Fe	274±19	300±20	267±5	270±20
Co	<6	(0.2)	<6	(0.18)
Ni	1.2±0.5	1.3±0.2	<0.8	–
Cu	11.5±1.0	12±1	192±4	193±10
Zn	25.3±2.1	25±3	134±2	130±10
Ga	<0.5	(0.08)	<0.5	–
Ge	<0.4	–	<0.4	–
As	10.1±0.8	10±2	<0.3	(0.055)
Se	<0.3	0.089±0.01	1.1±0.2	1.1±0.1
Br	9.0±0.5	(10)	8.8±0.4	–
Rb	11.5±0.6	12±1	18.4±0.4	18.3±1.0
Sr	36.3±1.3	(37)	<1	(0.14)
Y	<1	–	<1	–
Zr	<3	–	<3	–
Hg	<1	0.155±0.015	<0.8	0.016±0.002
Pb	40.7±3.0	45±3	<1	0.34±0.08
Th	<1	–	<1	–
U	<2	0.029±0.005	<2	(0.0008)

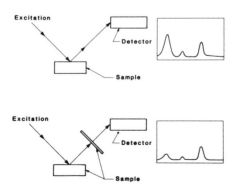

Figure 22.11 Direct measurement of the specimen matrix absorption coefficient, with which to compute calibration curves for the trace elements.

mass absorption coefficients for each element in the standard and unknown. Then the concentration in the unknown is given simply as:

$$C_u = C_s \ (I_u/I_s) \ (\mu_s/\mu_u)$$

where the subscripts u and s denote the unknown and standard, respectively. This method is mathematically simple, but requires difficult preparation of thin sections or pressed powders of the samples (the thickness must be great enough to cause significant absorption but not so great as to totally absorb the radiation, to give a reasonably precise determination of the mass absorption coefficient). The method is nonetheless sometimes used, for instance in the case of minor element analysis in the lunar rocks.

This and other unique methods can be tailored to particular analytical problems, by keeping in mind the fundamental assumptions regarding trace and minor element analysis: the calibration curve of intensity versus concentration will be linear, and the slope of the curve will vary inversely to matrix mass absorption coefficient. Altering the matrix absorption with heavy absorbers, or dilution of the sample into a low-absorbing glass to obtain constant and known calibration curves, are all accepted techniques.

Chapter 23

Modifying the Excitation Function

Thoughout the last few chapters, the X-ray intensity from each particular element in a complex sample was related to the total product (summed or integrated over all energies) of that element's absorption coefficient and the output of the X-ray tube used to excite the sample. By modifying the shape of the tube output, the relative yield of fluoresced X-rays from the various elements present can be changed dramatically, and may often be used to improve the analytical results for the elements of interest, either in terms of the precision attainable in a given length of time, or the detection limit that can be achieved.

The use of filters, secondary targets, and other modifiers has really only become practical with the advent of the energy dispersive spectrometer. Its high efficiency and low count rate tolerance allow rather low power levels to be used to excite the specimen. Consequently, since most techniques to modify the incoming excitation from an X-ray tube involve selecting some portion of its output and discarding the rest, it is still practical to use normal tubes and generators and have enough power left to keep the spectrometer counting busily. With WD systems, the same methods would be much less practical since the total system count rate would drop, in some cases by several orders of magnitude, and the loss in number of counts (and statistical precision) would degrade the results far more than the offsetting benefits of the specialized excitation.

It should be pointed out, in this context, that these modifications we shall describe are, almost without exception, ones that improve the ability to analyze some elements of interest at the expense of others. In some cases, the list of elements of interest is short and they are close together, and can be efficiently analyzed with the same modified excitation. In others, a series of separate conditions would need to be used, and the longer analysis time might offset (or more than offset) any benefit of the technique. For true unknowns, where it is necessary to check at least cursorily for the entire periodic table, continuum excitation with the broad spectrum of energies to excite most elements simultaneously is usually the best place to start.

Secondary Targets

The use of secondary targets is shown schematically in Figure 23.1. The continuum, or white radiation from the X-ray tube is used to excite another material, whose fluoresced X-rays (primarily characteristic X-rays of the element(s) present) then reach the sample to be analyzed.

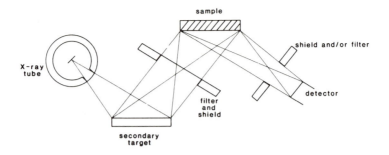

sample

X-ray
tube

shield and/or filter

detector

filter
and
shield

secondary
target

Figure 23.1 Diagram of secondary target used for specimen excitation in X-ray fluorescence analysis.

If the secondary target is a reasonably pure single element, the radiation striking the sample consists predominantly of one or a few characteristic lines, and in the common case of the K-α spectrum, the α line contains most of the intensity. This makes the integration of the product of $J(E)$ times $\mu(E)$ very simple, since it is not unreasonable to assume that $J(E)$ is zero except at the characteristic energy of the secondary target element. The fundamental parameters method, in which the integral equations are simultaneously solved, becomes rather straightforward once the integrals are removed, and it often becomes practical to perform "standardless" quantitative analysis using monoenergetic excitation from secondary targets, even for rather complex samples.

The excitation of the various elements in the sample depends on the intensity of the source and the element's mass absorption coefficient at the energy of the exciting line. If the line is below the element's absorption edge, there is no fluorescence, of course. If it is rather close to the edge, within a few keV., the excitation is strong and excellent sensitivity is obtained. If the exciting radiation is far above the absorption edge, the exponential decline in the absorption coefficient will result in a very low sensitivity, poor detection limits, and a low number of total counts. This will especially be evident if two elements of the same approximate concentration level in the same sample, are analyzed using monoenergetic radiation. The peaks, which would be roughly the same in size using continuum excitation, will be quite different. The element with an absorption edge close to the energy of the exciting line will give a peak in the spectrum much larger than the other element, farther away. Figure 23.2 illustrates this using a sample containing roughly equal concentrations of a broad range of elements (all K-line emitters) from calcium (Z=20, K-edge = 4.04 keV) to copper (Z=29, K-edge = 8.98 keV) excited by K-α X-rays from a germanium (Z=32, K-α = 9.88 keV) secondary target. The scattered exciting peak is also present in the spectrum, and we shall return to this aspect of monoenergetic excitation later on.

Note that the same sample, excited with continuum radiation, yields peaks that are roughly the same size, while for the secondary target excitation case, the lower energy elements have dramatically smaller peaks. This reminds us that a germanium secondary target would be fine for the analysis of elements such as iron and copper, but not for calcium or titanium. For a dedicated analysis job, we might be content to analyze just those few elements; in most cases we must find a way to equip our system to analyze a much broader spectrum of elements.

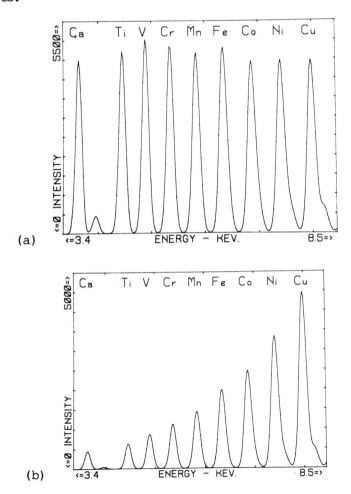

(a)

(b)

Figure 23.2 Spectra from a sample containing roughly equal amounts of K-line elements from Ca to Cu, excited by a) continuum, b) monoenergetic photons.

One solution, nearly universally employed, is to have a series of secondary targets available, often mounted on a wheel or drum and selectable externally or by the controlling computer. As each target is positioned, the appropriate tube settings of voltage and current are selected to strongly excite the characteristic lines of the target element, and then the portion of the spectrum containing those elements in the sample that are efficiently excited by the secondary target's X-rays is collected. Another, less flexible and much less common possibility is to use a composite secondary target (either a combination of several elements, perhaps pressed together as powders, or possibly a series of layers with the lowest energy emitter on the top for least absorption, and underlying layers of one or more higher energy secondary emitters). In the latter case, tailored targets for the particular needs of a particular type of sample and list of elements of interest can be produced which excite the various elements differently to yield the desired statistical precision for each in a single measurement time period. This is straightforward in principle: the high concentration elements need much less excitation to give an adequate number of X-rays than do the trace elements, and the amount of each secondary target element in the composite is varied to balance the excitation of the elements which it excites. In practice, such a specialized target can be quite difficult to design for an assorted list of elements, and once created it may not be useful for anything else.

But why should we want to only excite a few elements at a time? Why not just use the continuum from a typical X-ray tube and be done with it? Several reasons may arise. One is to minimize the excitation of a major element whose peak would otherwise dominate the spectrum (and the counting ability of the electronics). For instance, consider the analysis of trace elements such as molybdenum and zirconium in low alloy steels: the use of continuum would produce an enormous iron peak, accounting for over 90% of the measured pulses, under any conditions that would also excite the $K-\alpha$ lines of Mo and Zr. These peaks would be very small, and it would take tens or hundreds of seconds to obtain adequate statistics. But with a secondary target, perhaps silver ($K-\alpha$ = 22.1 kev, while the edges of Zr and Mo are at 18.0 and 20.0 respectively), we can strongly excite the minor elements of interest, and the iron peak will be greatly reduced, so that adequate statistical precision is obtained quickly. Figure 23.3 shows the appearance of spectra from a steel sample containing a few tenths percent Zr and Mo excited by continuum and a silver secondary target. The total analysis time was the same for each.

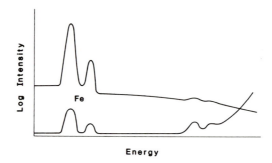

Figure 23.3 Comparison of spectra for low levels of Zr and Mo using continuum and monoenergetic excitation (logarithmic vertical scale).

In this figure, another change than the reduced iron peak is evident. The background in the spectrum is also reduced by the use of a secondary target's monoenergetic excitation. Remember that the spectrum background is primarily due to the scattering of radiation from the sample to the detector. In the case of secondary target excitation, the spectrum of radiation reaching the sample contains very little intensity at any energy except the target element's characteristic lines. The scattering of continuum originating from the tube, first from the secondary target and then again from the sample, to reach the detector, is greatly reduced. This is especially true since most targets are made from comparatively high atomic number materials (Ag, for instance, is Z=47) which scatter less intensely than low atomic number samples.

It is the reduction of spectrum background that permits the peaks from minor or trace elements in the sample to be readily detected. Detection limits can often be improved by an order of magnitude or more for elements which can be strongly excited by a secondary target, and under whose peaks there is very little scattered background intensity. Incidentally, this background radiation also must be counted by the system electronics, so it might be supposed that a further improvement in the throughput of the X-rays of interest would result from the elimination of the background. This is not really so, since the monoenergetic excitation line is still scattered to the detector and counted, and at least to a rough approximation will have as much intensity as the continuum. Because the monoenergetic line is always at a higher energy than the lines of the elements being analyzed, it is sometimes possible to discriminate against these pulses either electronically, so that they need not be fully processed and counted, or by insertion of a filter to prevent their reaching the detector at all.

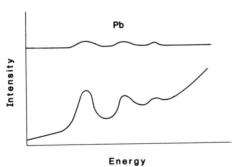

Figure 23.4 Trace element analysis (lead in organic matrix) requires reduction of scattered background by the use of monoenergetic excitation (logarithmic vertical scale).

Reduction of scattered continuum background is particularly useful for the analysis of trace elements in low atomic number matrices such as organic materials (petroleum products including oils and plastics, foods, and so on) and minerals. The example shown in Figure 23.4 illustrates the improvement in detectability of trace amounts (a few parts per million) of lead in feed grains. With continuum excitation, the peaks are barely visible (and have poor statistical precision) above background; with monoenergetic excitation using a zirconium secondary target (K-α = 15.75 keV while Pb LIII absorption edge = 13.04 keV) the peak to background ratio is improved. In the same total analysis time, the detection limit and analytical precision are greatly improved. Remember, of course, that it is necessary to increase the tube power to get this result, and the simultaneous analysis of potassium, for instance, is no longer possible.

To give general coverage for any element which might need to be analyzed, we should require a fairly broad selection of secondary targets. We want an exciting energy that is between about 1.1 and 1.3 times the absorption edge energy for the analyte. If it is closer to the edge than that, there may be interference between the scattered characteristic line from the target and the line we are trying to measure. If it is farther away, the efficiency of excitation will suffer. At very low energies, corresponding to the K-lines from elements up to about sulfur, it is difficult to find suitable secondary targets that will allow the analysis of more than one or two elements at a time. Consequently, in this energy range, it is more common to use continuum radiation directly from an X-ray tube. However, at higher energies, targets such those shown in the table below can be selected. The emission energy of the target and the absorption edge energies of the analytes are listed. Intevening K-line elements, and L-line elements with energies in the same range would also be covered:

Target element	Typical analyte elements	
Ti (4.51)	K (3.61)	−Ca (4.04)
Fe (6.40)	Ti (4.96)	−Cr (5.98)
Cu (8.04)	Mn (6.54)	−Co (7.71)
Ge (11.1)	Co (7.71)	−Zn (9.66)
Zr (15.75)	As (11.9)	−Rb (15.2)
Mo (17.44)	Kr (13.5)	−Sr (16.1)
Ag (22.10)	Sr (16.1)	−Mo (20.0)

There is something of a gap in the range around 13 keV, which corresponds to the K-line of bromine (not a good choice for a secondary target). While the requirements for a good secondary target element are much less stringent than those for an X-ray tube target (no need for electrical conductivity, little problem with heat conduction, no need to be inside a vacuum, unless the entire spectrometer is evacuated −and for these higher energies it need not be), a gas like bromine is still clearly unsuitable. Rubidium metal (K-α = 15.2) might be used, or perhaps an L-emitter like uranium (L-α = 13.61). The L-line spectra are less desirable for secondary target excitation because a smaller fraction of the generated radiation is in the single most intense line. The other higher energy lines are rarely useful for excitation because the elements they excite have emission lines which will be interfered with by the scattered lines from the target element. Other selection of target elements could of course be made, and would even be encouraged if particular analyte elements were known ahead of time to be of interest. The list is shown primarily to emphasize the need for a variety or secondary target materials to obtain optimum coverage.

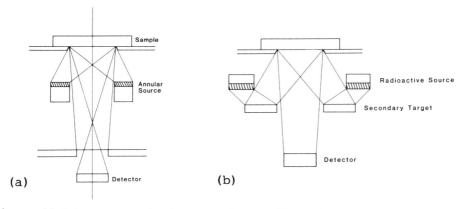

(a)   (b)

Figure 23.5 Arrangement of an annular radioactive source for X-ray fluorescence analysis, either (a) exciting the specimen directly or (b) with a secondary target.

Radioactive Sources

Before leaving the subject of monoenergetic excitation, it is useful to point out that secondary targets excited by X-ray tubes are not the only simple source for such radiation. Radioisotopes are widely available in small, encapsulated and easily handled sources, which emit either γ- or X-rays at a variety of energies. The simplest type would be an emitter such as cobalt-57 which emits, among other things, a γ ray at 14.1 keV. This could be used for excitation of a sample directly. There is no difference whatever between γ and X-ray photons (both are electromagnetic radiation whose characteristics are fully described by their energy or wavelength), except for their origins (X-rays in processes involving inner shell electrons, γ rays in processes within the atomic nucleus). Many radioisotopes emit X-rays from the excitation by γ rays of the daughter element in their decay. For instance, a cadmium-109 source emits primarily silver K X-rays, indistinguishable from those from a bulk silver secondary target. There are many such isotopes available commercially, but they do not offer complete energy range coverage to efficiently excite the elements of probable analytical interest (Rhodes, 1966; Kneip, 1972).

Another approach is to use a secondary target to produce X-rays of the energy desired, but to excite it with a radioisotope source. This can often be accomplished within an extremely small volume, by using annular sources and targets as shown in Figure 23.5.

This allows the source and detector to be placed very close to the sample, so that low power levels (in this case, low source strengths) can be employed and still yield good count rates and analytical precision. The chief advantage of such systems is their lack of a separate high-power tube and generator, and the extreme stability of the source. Even for relatively short half-life isotopes such as cadmium (which requires correction for its dropping intensity every week), there is no problem with short term fluctuations in the output as can occur with a tube. The development of compact, even portable XRF systems using radioisotope sources has opened a rather specialized market for on-site or even down-borehole analyzers. One important difference that may arise between analysis using X-ray tubes and isotope sources is that the former can usually be operated at voltages up to perhaps 50 to 60 keV., exciting reasonable intensities from target elements (either the tube target or a secondary target) at energies up to about half that value. Consequently, monoenergetic excitation using X-ray tubes is typically practical for K-lines of elements up to about atomic number 40-50 (and of course for the L lines of the entire periodic table). With radioisotope sources, much higher energies are available, as shown in the following table of some of the more common sources. Whether used directly or used to fluoresce secondary targets to obtain more desireable energies for exciting specific elements, this means that K shell excitation may be possible for much higher atomic number elements, such as gold (K-α = 68.2 keV) or uranium (K-α = 97.1 keV.), both of which are of considerable interest in mineral exploration.

Common Radioisotope Sources

Nuclide	Half-Life	Photon Energies	Elements Excited
Fe-55	2.7 years	5.9 keV	V and below
Cd-109	453 days	22.1 (Ag Ka)	Cu-Mo (K shell)
Gd-153	242 days	41,70,97,103 keV	Mo-Ce (K shell)
Am-241	458 years	59.6 keV and	
		14-40 keV Np L X-rays	Sn-Tm (K-shell)
Co-57	267 days	14,122,136 keV	Ta-U (K-shell)

(Note: in addition to the elements listed, excitation of elements with L shell absorption edges in the same energy range is also practical. Also, particularly for the Am-241 source, a variety of compact secondary target assemblies are available which produce essentially monoenergetic exciting radiation at many other energies, for specific applications)

The advantage of using the very high energy K lines for analysis is that they are absorbed but little in passing through a typical matrix of oxide minerals. Consequently, the depth of analysis of very large samples and sampling error in quite heterogeneous materials is reduced. Also, the matrix effects which produce non-linear calibration curves are reduced when absorption is reduced, and simple linear calibration curves or addition techniques become useable. On the other hand, the high energy X-rays are not efficiently detected by typical Si(Li) detectors (the high energy radiation passes through, perhaps fluorescing the supporting hardware, but not producing identifiable pulses). Germanium detectors, either of the lithium compensated type or, more recently of high purity germanium that is sufficiently defect free that this is not required, are preferred for this energy range. They have higher absorption coefficients for these high energies, and in suitable thicknesses of a few millimeters can give adequate detection efficiency for high energy K-α lines. Figure 23.6 shows a typical spectrum for gold in a silicate matrix, excited with the 122 keV. gamma ray from cobalt-57. Using other energies for excitation, for instance with a secondary target arrangement, and analysis times of 5 minutes, detection limits for gold of 1-2 parts per million in a measured volume of 25 cubic centimeters can be achieved. Such a system can furthermore be relatively transportable, requiring only generator power for the counting electronics, and a replenishable supply of liquid nitrogen to cool the detector. The use of mercuric iodide room temperature detectors could remove even the latter requirement. Interest in such analytical tools continues to bloom in the minerals exploration field.

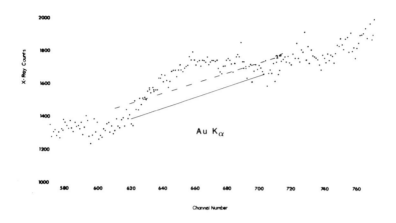

Figure 23.6 Spectrum showing gold K peaks measured from a sample containing 17 ppm. gold in a silicate matrix, with high energy radioisotope excitation.

The basic principles of excitation are not different for these systems than for other secondary target or monoenergetic excitation systems, however, and beyond the obvious challenge of finding the proper source and target, and constructing the necessary hardware to hold everything in position, there is no further need to discuss them.

Filters

A very different approach to modifying the exciting spectrum is to place filters between the X-ray tube and the sample, to absorb some of the radiation (Vane, 1980). This is a distinct technique from the other use of filters, placing them between the sample and the detector. In this latter case, the purpose is to remove some fraction of the spectrum so that it need not be counted, thus improving the performance of the count-rate-limited ED system electronics for the elements of interest. Such cases are comparatively unusual, requiring that a major elemental emission line be rather far away in energy from the lines of interest (in which case, it may be easier not to excite the element either by using a monoenergetic excitation source or by changing the voltage on the X-ray tube). Placing a rather special filter, consisting of the sample itself, in the position between the sample and the detector, was also mentioned before as a means of measuring the matrix absorption coefficients directly.

When a filter is placed between the tube (or other source of primary radiation) and the sample, it can only absorb X-rays, more at one energy than another, to modify the energy distribution of the exciting radiation. As in the case of secondary targets, the advantage this produces analytically may be either from the relative increase of a particular energy of photons which strongly excite an element of interest, or from the removal of radiation which would scatter from the sample to become background in the spectrum. The selection of filters to accomplish this is something of an art, since the element(s) used in the filter contribute strong absorption at and above their respective edges, but also general absorption at all energies based on their thickness. Often, a particular filter will be designed with only one of these two functions in mind, and will be referred to as either an "edge" filter (to selectively remove a portion of the tube spectrum just above the edge) or a "white" filter in which the thickness and mass of the material cut off all radiation below a certain energy.

As an example of the latter, consider the need to analyze a low concentration of a low energy element such as phosphorus (K-α = 2.01 keV) in a low atomic number matrix (such as oil). The scattered radiation from this matrix will impede our ability to see the small elemental peak, and it could be reduced by cutting off the radiation striking the sample at that energy. If we can reduce the intensity of the exciting spectrum at 2 keV, more than we reduce it at 2.5 or 3 keV (which is the energy of the photons that do most of the exciting of the phosphorus atoms in the sample), we can improve the peak -to - background ratio and consequently the detection limit. Placing a thin foil of any material in the path of the beam will absorb some of the X-rays, and because we know that except for the contribution of the absorption edges, the mass absorption coefficient rises as energy drops, and indeed that the rise is roughly proportional to $1/E^{2.8}$, we see that this can be accomplished using a filter made from almost anything. A foil of nickel, or aluminum, or even cellulose, would work.

The next step is to decide on the optimal thickness. We want to improve the ratio of peak (excited by X-rays at, say, 2.5 keV which we therefore do not want to absorb) to background (scattered 2 keV X-rays, which we do want to absorb). The thicker the foil, the greater the total absorption, and the lower the total count that will be obtained in a given time. The detection limit is improved in proportion to the product of count rate and peak-to-background ratio, and so we can resort to Beer's law $I/I_o = \exp(-\mu(E)\rho t)$ where μ is the mass absorption coefficient for the material, at each of the two energies, and pt is the mass thickness of the filter, to find the thickness which gives the best result.

It is often possible to do a better job of reducing background under a particular trace element peak by using a filter of an element with an absorption edge just below the peak to be measured. The example in Figure 23.7 shows the selective reduction of background under the same trace lead peak shown before with secondary excitation. In this case, a zinc filter is used to remove the scattered background underneath the peak. The zinc filter also reduces somewhat the intensity of the radiation that can excite the lead, but particularly for L line spectra, these energies are rather far above the emission line (the lead LIII edge is at 13.04 keV, the zinc absorption edge at 9.66, and the lead peak at 10.55 keV) and the reduction of peak intensity is minor.

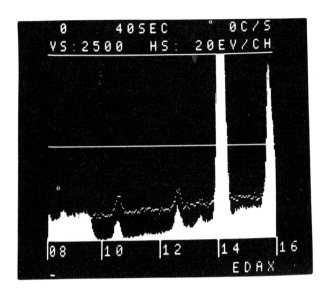

Figure 23.7 Selective reduction of background under trace lead peak using a zinc filter.

Another use of filters between the X-ray tube and the sample is to pass with minimal absorption the characteristic line of the X-ray tube, while reducing the continuum background just below it in energy (where the characteristic peaks of the elements which can be efficiently excited by the tube line will lie). This is most often accomplished by using a filter material identical to the tube anode, for instance a Mo filter for a Mo tube, since the mass absorption coefficient of an element for its own radiation is rather low (the emission lines are not high enough in energy to excite the element, of course). The filter will absorb continuum radiation at lower energies, not selectively as in the "edge" application described before, but simply with a mass absorption coefficient that increases exponentially with falling energy. This produces an exciting radiation spectrum such as that shown in Figure 23.8, which is far from monoenergetic, but nevertheless an improvement over the broad continuum distribution for purposes of detecting small peaks in the region below the characteristic line. Whether this is of any practical use depends, of course, on whether the choice of tube anode material was made to permit selective excitation of the elements of interest by its characteristic line. Usually, these situations arise only fortuitously. The use of a filter in this way is simply a poor approximation of the more general and powerful method of having several selectable secondary targets, which can be selected to cover a broader range of possible elements of interest. It is worth noting in passing that this scheme is very inexpensive to implement, compared to secondary target systems, and that several manufacturers have not only offered such systems but have made extravagant claims for the purity of the resulting "monoenergetic" radiation, implying that the filter acts itself as a secondary target. Simple calculations show that the intensity of characteristic X-rays emitted from the thin filter material, excited only by the continuum above the element's absorption edge energy (and hence above the characteristic emission lines) is negligible, amounting to less than the reduction in the characteristic line intensity by absorption.

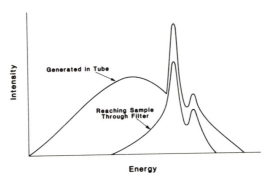

Figure 23.8 Shape of excitation spectrum using a filter of the same element as the tube anode.

More useful than the preceding application of filters is one in which the tube characteristic line is removed by a filter. This can be important if the tube anode element produces scattered peaks in the measured spectrum that interfere with minor peaks from elements of interest, including the anode element itself and any others whose peaks lie in the same region of the energy spectrum (including the possibility of interference with the Compton scattered peaks). This can be accomplished by inserting a filter made of an element with an absorption edge just below the tube element's emission energy (for instance, for a W tube the principal emission energy is the L-α-1 line at 8.40 keV, and this is strongly absorbed by a nickel filter with a K absorption edge energy of 8.33 keV.).

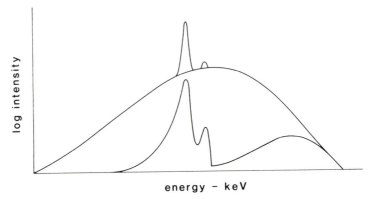

Figure 23.9 Shape of exciting spectrum using a filter to remove the tube characteristic line but pass higher energy continuum.

When the tube characteristic line is removed in this way, the spectrum of energies available to excite the spectrum has a "hole" in it where elements will not be strongly excited (fortunately there is still some higher energy radiation that passes through the filter, as shown in Figure 23.9). The ability to analyze trace and minor elements with peaks in this region is inferior to what could be achieved with the use of a different tube anode material together, but vastly improved over the original, unfiltered case. This is perhaps the lowest cost way to make systems employing only a single X-ray tube which directly excites the specimen, into a reasonably complete tool for analysis of all elements.

In addition to the typical X-ray tube configuration which has been assumed thus far, in which a beam of electrons strikes a bulk anode, and the X-rays pass through a thin beryllium window to strike the specimen, several other more specialized geometries have been used for particular purposes. One, which has no window, allows very low photon energies to reach the sample to more strongly excite the low energy elements. In this case, magnetic or electical fields may be employed to direct the electron beam onto the anode and prevent the backscattered electrons from reaching the specimen (where they would generate Bremsstrahlung background, and possibly cause charging or damage). These designs have never become popular, because ultralight element analysis using X-rays is so difficult anyway (the shallow depth of excitation places extreme burdens on specimen preparation of a representative surface, and the low fluorescent yields result in low measured intensities). Other analytical techniques are usually preferred for these elements, including analysis of the Auger electrons that are emitted from those atoms which return to the ground state after being excited, without emitting an X-ray photon.

Another variation of tube design is to use a thin foil anode which is itself the window through which the X-rays exit from the tube. This geometry allows the source to be placed very close to the sample, and consequently allows even lower power tubes to be used. The thickness of the anode/window is critical, since it must support a one-atmosphere pressure difference and yet be thin enough for the X-rays generated by the electron beam striking one side of the foil (and as we have seen before, penetrating to a depth of a few micrometers at most) to suffer minimal absorption. These is still the same effect as in placing an external filter between the tube and sample, of the same material as the anode, so that the low energy portion of the spectrum is especially attenuated. These tubes offer little flexibility and are now infrequently used.

Most other peculiarities of tube design have been concerned with the cathode and grid arrangement to produce highly stable electron sources that can be pulsed rapidly on and off. The utility of this was previously described as a means to reduce the statistical randomness with which X-rays reach the detector, and allow more pulses to be processed through the main amplifie. Increases of 2.5 to 3 times in count rate can be achieved using pulsed tubes, and for the comparatively low power levels needed for energy dispersive analysis, such tubes and control grids and circuits and be practically made. Most energy dispersive systems now incorporate such tubes, because of the recognized need for doing everyting possible to increase the total system countrate capability (a principal acknowledged limitation of ED systems).

Polarized X-rays

Another approach to modifying the exciting spectrum utilizes the principles of polarization to reduce scattered backround in the measured spectrum (Ryon, 1982). It is well known that polarized sunglasses reduce glare, for instance due to the scattering of sunlight from the surface of water. This happens because in the scattering process, the component of the light which has its polarization vector parallel to the water surface is scattered, but the component with its polarization vector perpendicular to the surface is not. Consequently, placing a polarizing filter in front of the eye to transmit only light with a vertical polarization results in a "glare-free" image. Since X-rays are electromagetic radiation, too, it should come as no surprise to find that in scattering of X-rays from the sample surface, the same kind of polarization occurs. There are no handy polarizing filters or "sunglasses" for X-rays, to place in front of the detector. Instead, we can arrange to excite the specimen with polarized X-rays which have a vertical polarization vector. These will not be scattered, and hence only the radiation generated in the sample will be measured. The polarization of the incident X-rays is produced by scattering of that radiation from another surface. Figure 23.10 shows the geometric arrangement of the tube, polarizing scatterer, sample, and detector.

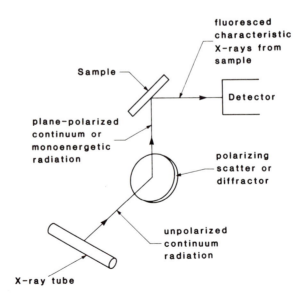

Figure 23.10 Arrangement of source, scatterer (polarizer), sample and detector to investigate use of polarized X-rays for excitation.

Note that this figure looks a lot like the arrangement for secondary target excitation, except that the plane containing the tube, scatterer and sample is perpendicular to the plane containing the scatterer, sample and detector, It is this out-of-plane arrangement that produces the reduction in background due to scattering. All of the angles are 90 degrees, which minimizes incoherent scattering. The continuum spectrum from the X-rays tube strikes the scatterer, typically a low atomic number material, and produces a spectrum that strikes the sample that has essentially the same energy distribution but is now polarized. When these X-rays reach the sample, they produce excitation as usual, but with very little further scattering toward the detector. The result is that only the characteristic lines from elements excited in the sample are detected, and consequently the detection limits are greatly improved. The spectrometer system could, of course, also be used for conventional secondary target excitation by replacing the scatterer with the secondary target, but the converse is not true since existing secondary target excitation systems do not have the necessary perpendicular arrangement of components.

The process of scattering to produce polarized X-rays is very inefficient, so only a small fraction to the total tube output reaches the sample. Using conventional 2-3 kilowatt tubes (as in conventional wavelength dispersive systems), the count rate at the detector is quite low, well within the the capability of an energy dispersive system. The degree of polarization is controlled by limiting the divergence angle of the beams, since at angles other than 90 degrees some scatter occurs with polarization vector normal to the surface. With real systems, the divergence angles are finite and imperfect polarization results. The consequence is that there is some scattered spectrum background, although greatly reduced from the unpolarized case. The usual method for optimization is to set the divergence angles so that the total system count rate is just tolerable, which gives the greatest possible intensities for the characteristic lines along with good peak to background ratios by reducing scattered background. The product of peak to background ratio and net elemental intensity determines detection limits.

Although at present this is an experimental technique, it holds great promise. Table 23.1 compares detection limits on elements in several NBS trace element standards using secondary target and polarized excitation, and shows that the polarized results are as good or better, while at the same time covering a much broader range of elements and energies than could be excited simultaneously using a single secondary target. Further development work, including methods for increasing the degree of polarization without reducing total system count rate, should make this method very attractive for trace multi-element analysis. One novel approach is to fabricate an ellipsoidal scatterer, with the tube at one focus and the sample at the other, to increase the geometrical efficiency of the system and utilize a much higher fraction of the tube's output power.

Table 23.1 Trace element analysis capabilities using polarized excitation (Ryon, 1982).

<table>
<tr><td colspan="4" align="center">Detection Limits
ppm. $(3-\sigma)$ in NBS Orchard Leaves
comparing polarized and secondary excitation</td></tr>
<tr><td colspan="4" align="center">(for each element, the same generator voltage
and current were used for all excitation methods)</td></tr>
</table>

Element (K-lines)	Secondary Target	Bragg * Polarized	Barkla ** Polarized
	Ti	CuKα –Ta(002)	B$_4$C scatterer
S	50	120	170
Cl	18	50	65
K	6.2	7.3	14
Ca	5.8	5.4	11
	Y		
Fe	1.9	1.9	.84
Br	0.3	—	0.3
Rb	—	—	0.3
Sr	—	—	0.3

* by diffraction at 90 degrees
* * by scattering at 90 degrees

Chapter 24

Other Ways to Excite the Sample

The methods described thus far for producing inner shell excitation of atoms in a sample of interest, so that characteristic X-rays are produced and can be measured, are by no means the only possible or practical ones. Any means of delivering enough energy to overcome the binding energy of the electron will work. In addition to electrons and X-ray photons, other particle or photon sources can be used. In fact, the use of γ ray photons from radioactive sources has already been mentioned, and as it is identical in terms of the analytical results that are produced, with monoenergetic X-ray excitation from a secondary target, it will not be discussed further here. Since any form of photon excitation, regardless of the source of the photons, will be indistinguishable in terms of the interaction of the photons with the atoms of the sample, there is in fact no need to separately consider different sources of photons except as their energy distribution may be altered, as was discussed in the preceding chapter.

Particle excitation, however, can be accomplished with many other types of particles than simple electrons. The ones we shall consider here are protons and alphas, although there have been at least occasional uses of heavier atomic nuclei. For all charged particle methods, there is need to consider the background produced by Bremsstrahlung, not present with photon excitation. However, the very different masses of the various particles strongly affects the energy distribution of this background. Also, the depth of penetration of charged particles, even quite high energy ones, is much less than for photons. We have already seen that the typical depth of excitation for electrons is on the order of a micrometer, while for X-ray fluorescence it can easily exceed a centimeter for some combinations of elements and sample matrices. This is another factor to be considered, as it has important consequences for what types of samples and preparation can be profitably employed with the various types of charged particles.

Proton Induced X-ray Emission (PIXE)

The use of protons for excitation of characteristic X-rays, which are subsequently detected with energy dispersive, usually Si(Li) detectors, has a history older than the use of electrons. The limited amount of work that has been performed using these techniques, and the comparatively smaller literature that exists, is due primarily to the fact that producing high energy beams of protons is much more expensive than the typical scanning electron microscope. Beams of protons from accelerators or large Van de Graaf generators, with currents in the nanoampere range and voltages of typcally 2-4 MeV. are available at only a relatively few locations throughout the world. The analytical use of these beams for X-ray analysis of samples is rarely a major source of justification for the initial construction of such a facility.

Nevertheless, the availability of such a source permits some excellent analytical work to be accomplished. At these high energies, the protons have a very high cross section for inner shell excitation of atoms in the sample. The cross section, or probability of excitation, is greatest for low atomic number materials and falls as Z increases (the opposite behavior to the variation of the photoelectric cross section, or mass absorption coefficient for X-rays, which increases with atomic number). Proton Induced X-ray Excitation (PIXE) is therefore capable of good sensitivity especially for the lighter elements, with somewhat poorer detection limits for heavier elements.

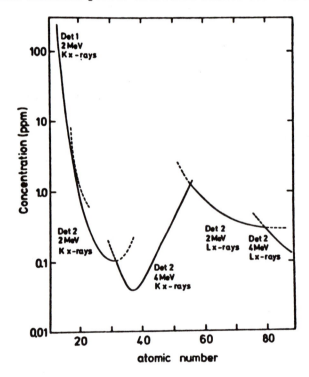

Figure 24.1 Detection limits using PIXE, compared to secondary target XRF (Scheer, 1977).

The production of Bremsstrahlung by the deceleration of protons in the electrostatic field around the nuclei of atoms in the sample will produce spectrum background, just as in the case of electron excitation. However, the protons are much heavier than electrons and so the amount of deflection they can undergo in a single interaction is much less. Consequently, the energy loss is small and the energy distribution of the generated continuum background shows a peak at very low energy (usually below the detection level of the spectrometer) and a rapid fall off with rising energy. For much of the spectrum of interest, the background is extremely small; where it is high, the large ionization cross section for the low atomic number elements, which produce low energy characteristic emission peaks that appear on top of the background, gives very large peaks which are still easily detected.

The consequence of these factors is that PIXE can often produce detection limits as much as an order of magnitude better than monoenergetic (eg. secondary target) X-ray photon excitation, and at the same time cover a wider energy or elemental range. This is illustrated in Figure 24.1, showing estimated detection limits for the same samples using the two techniques.

However, there are some important limitations on the nature of the samples that can be suitably analyzed by this method to obtain such favorable results. If the specimen is too thick, the deceleration of the protons gives rise to secondary, but rather high energy electrons which themselves are capable of producing Bremsstrahlung radiation as they penetrate through the material. This background in the measured spectrum seriously degrades the detection limits. Consequently, it is important to have the specimens be as thin as possible. This also reduces Compton scattering in the sample.

Very thin substrates such as Formvar films, on which the residue from evaporated liquid samples remains, make excellent samples for the analysis of trace levels of elements in blood serum, urine and other biological fluids, for instance to detect trace levels of toxic environmental elements. Such thin substrates are very fragile and easily damaged by heating or charging by the beam. Thicker Mylar films, or Nuclepore filters, are used for their greater strength in spite of the greater background they produce. These are especially used for the analysis of filtrates from liquids or air, the latter for analysis of environmental air samples. Sandwiching of fine particle material between layers of Mylar, or pressing thin pellets of samples such as minerals, in which about 10% graphite is incorporated as a binder and electrical conductor, makes it possible to analyze finely ground bulk materials as well. However, as the thickness increases, so does the background in the spectrum.

Figure 24.2 shows the typical appearance of a PIXE spectrum, with its pronounced low end background. Note that the vertical scale is logarithmic, and that the trace element peaks which represent part-per-million level concentrations are nonetheless quite evident above background.

Spectra like this are usually processed by simple linear least squares fitting of component spectra representing the background from a clean substrate, and pure elemental characteristic spectra measured from standards. The extremely thin samples have negligible absorption or other non-linear effects on the X-ray emission, and so it is a good approximation to represent the total measured spectrum as a linear combination of the various components. This also means that calibration curves of intensity versus concentration should be linear, so calibration or standardization can be very straightforward.

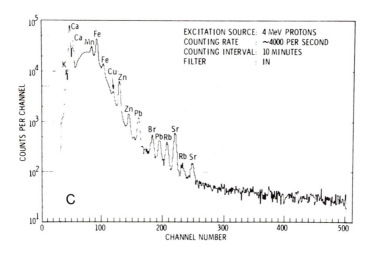

Figure 24.2 PIXE spectrum from trace element specimen (Cooper, 1973).

Alpha Particle Excitation

The use of high energy α particles for X-ray excitation is somewhat similar to protons, with of course the particle mass being even heavier. The problem of spectrum background from thicker samples is generally even worse, and as illustrated in Figure 24.3, Cooper has shown that the practical detection limits that can be achieved are not as good with alphas as with protons (indeed, he also showed that in many cases, the proper choice of monoenergetic X-ray photons from secondary targets gives as good results).

Another use of α particles, however, has been developed by Musket (1977) into a commercially useful instrument. This takes advantage of the fact that heavy charged particles like alphas penetrate only a short distance into solids, and that the generated X-rays (which as for protons, are especially strongly excited from low atomic number materials for which the cross section is large) can escape to be detected with little absorption. He used for excitation a radioactive source (Cm-244) which emits 5.8 MeV α particles, arranged as an annular source around a Si(Li) detector which could be brought very close to the sample with no intervening window. The system thus required a fairly good vacuum, and even then for some samples which fluoresced visible light, a thin window was required to prevent the light from reaching the detector. However, it became practical to detect the ultralow energy photons from elements such as carbon and oxygen, fluoresced from surface layers of the sample by the α particles, with extremely low background levels as compared to electron excitation.

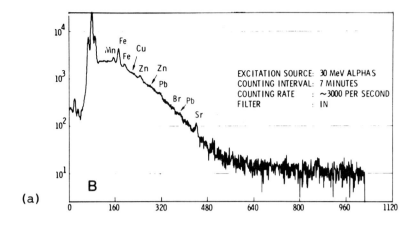

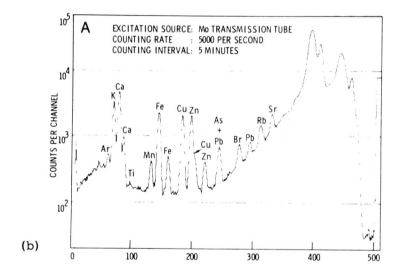

Figure 24.3 Spectra from trace element specimen, using (a) particle excitation and (b) conventional secondary target XRF (Cooper, 1973).

Examples of the use of this technique include the quantitative measurement of oxide film thickness on metals, and of carbon in surface layers of a steel. In the first case, the oxygen intensity from an unknown (anodized tantalum, as shown in Figure 24.4) was linearly related to thickness by direct comparison to the intensity from an oxidized silicon standard. The extreme thinness of the oxidized layer and the shallow depth of excitation produced by the alphas allowed absorption of the low energy oxygen X-rays to be ignored, which would not have been possible with electron beam excitation (efforts to measure oxide film thickness with the electron microprobe, even using wavelength dispersive spectrometers, usually require extensive absorption corrections).

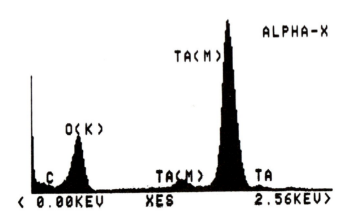

ANODIZED TA #1   ML   Z=73 TA
PR=   720S   720SEC   0 INT
V=4096 H=10KEV 1:2Q AQ=10KEV 1Q

ALPHA-X

TA(M)

O(K)

C   TA(M)   TA

< 0.00KEV   XES   2.56KEV>

Figure 24.4 Spectrum from surface of anodized tantalum excited with α particles (Musket, 1981).

A similar technique to measure carbon surface concentration on contaminated steels showed that layers a few tens of angstroms thick could be detected. Combination of measurements such as these with ion sputtering to remove surface layers can also permit depth profiling. It has already been pointed out that detector and electronic noise characteristics make it difficult to measure these low energy X-rays, and that the fluorescence yield is low as compared to Auger electron production. However, the use of α excitation as compared to electron beams offers the advantages of much higher ionization cross sections and better detection limits, shallow depth of excitation which minimizes absorption effects, and above all a practical way to utilize very clean vacuum systems so that the absorbing window in front of the X-ray detector can be removed. On the other hand, the ability to focus the electron beam permits measurements on a lateral scale of micrometers, for light or heavy elements, while any scheme using alphas from an isotope source will be limited to averaging over several millimeters. Proton beams are similarly usually used to uniformly excite regions with a diameter of about a centimeter.

Chapter 25
Sorting, Tagging, and Matching

By virtue of its ability to provide analytical information on many elements simultaneously (provided that either a multi-crystal wavelength spectrometer, or an energy dispersive one with suitable counting electronics is used), X-ray analysis has often been applied to problems involving recognition, identification or verification of composition based on a "signature" of many elements at once. Actually, the range of applications cover some very different situations, which we may summarize as follows:

1) confirmation that one or more elements are present in a predetermined range of concentration, as for on-line quality control checking ("go / no-go")

2) determination of the presence or absence of several of a predetermined list of elements, often intentionally added to a material, to provide a recognizeable and identifying tag ("tagging").

3) comparison of a measured pattern of elemental intensities to a series of stored patterns to identify which type of material is present ("sorting").

Even within these categories, there are some important variations depending on the particular application.

Consider first the routine go/no-go check of finished parts (typically, although it could be applied to raw materials as well) to be sure that one or more elements are present within a preset acceptance range. The test is usually made continuously, so that many parts or a significant volume of material is monitored. The count rate for the element(s) of interest may indeed be fed back to a control point to close the loop. The primary difficulty in this type of application is the variation in intensity due to simple counting statistics, which imposes a limit on how long an integration period (and thus how much material) is needed to obtain adequate precision, and the inherent assumption that the intensity variation reflects only a change in the parameter being controlled, whether it is the composition of the elements, thickness of a coating, or some other similar variable. Most often, a simple measurement of a single element is also likely to show variation due to the sample surface roughness, size or orientation, or to minor changes in position relative to the detector and source of excitation. The usual solution to this problem is to use ratios to approximately cancel out these geometric effects.

As a representative example, consider the need to control the thickness and composition of a brass coating applied to steel wire used for tread reinforcement in automobile tires. The wire passes at high speed beneath an energy dispersive Si(Li) detector and an X-ray tube or a suitable radioactive isotope source such as Cd-109, which excites the copper and zinc in the coating and the iron in the substrate. The ratio of copper to zinc intensity is a simple function of the coating composition, and the ratio of copper to iron a simple function of coating thickness. It is not important that these functions be linear, or even fully quantified by many measurements on standards. Rather, it is only necessary to measure ratios at the upper and lower limits of acceptable product specifications, and then adding appropriate factors of conservatancy, and considering the statistical precision needed and the time required to obtain it (which in this particular example is about 20 seconds, much less than the response time of the wire coating mill), to set upper and lower limits which will produce an appropriate signal to the equipment or operator.

In some mineral beneficiation processes, it is similarly necessary to monitor the concentration of one or more elements in either the primary product or in the tailings. In this case, it is often easier to use the intensity of the scattered peak from the exciting source (often a radioactive isotope) to compensate for variation in density of a flowing slurry, for example, than to use elemental ratios. This method has been used with compact immersible probes that can be inserted into flowing slurry streams to analyze for several elements of interest simultaneously. Alternative schemes bring the flow to a central location where special flow cells can be analyzed by the X-ray equipment. In either case, the need for special hardware to handle the flowing material or protect the source and detector imposes some substantial design constraints, and the presence of Mylar films to protect the detector, and so forth, coupled with the use of radioactive sources for excitation, usually makes light element analysis difficult or impossible. Monitoring of transition metals and heavier elements is straightforward, however.

Tagging

In the preceding examples, the elements analyzed were all present naturally in the sample, and it was a change in the relative intensity (to each other or to a scattered peak) that was related to the concentration being controlled. We sometimes simply need to know if an element is present or absent. This could be for elements really in the sample (for instance, monitors have been designed to detect lead in gasoline, toxic elements in feed grains, or lead in paint), but in many situations, the elements detected are not really in the sample itself at all, but in a separate "tag" added or attached solely for X-ray detection. In either case, the problem becomes one of deciding whether or not an element is present, in a short time.

This is a statistical problem. The number of counts in a given energy (or wavelength) band in a given time will, on the average, be greater when the tag element is present than when it is absent. However, it will not be zero when the element is absent (there is a finite background level due to scattering), and it will not be constant when the element is present. Even if the tag always contains the same amount of the element, variations due to the placement of the tag on the sample (several examples will be described shortly), and of course X-ray counting statistics, must be taken into account. Considering only the latter for the moment, the number of counts obtained in a short time (somehow these measurements are always supposed to take as short a time as possible, so that lots of samples can be examined) will vary following a Poisson probability distribution. If the mean numbers of counts per unit time that would be obtained with the element absent and with it present are B and P respectively, the situation is as shown in Figure 25.1.

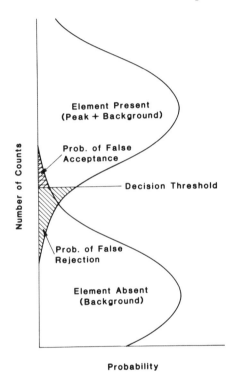

Figure 25.1 For tagging, a decision threshold must be placed at a number of counts that gives acceptable rates of false acceptance and rejection.

If the mean intensities become large, these distributions become indistinguishable from the normal or Gaussian distribution. Note that the probability distribution is such that it is possible, even if unlikely, to occasionally count a very small number of X-rays when the element is present, or a very large number when it is absent. These are the probabilities of false negative and false positive indications, respectively. The selection of a particular decision threshold, or number of counts at which the element will be considered to be present, must be based on a consideration of these probabilities. For instance, if the tags are being used to keep track of part identification, it may be comparatively inexpensive to occasionally discard a part which was really correct, but potentially very expensive to accept one in error which was really wrong. These probabilities are the areas under the tails of the probability distributions which extend beyond the decision threshold. It is possible to make one much smaller than the other, by suitable selection of the decision threshold.

When the additional complications (and their respective probability distributions, which may not be easy to estimate) of permitted variation in the concentration of the element in the tag, or the tag size, and its placement with respect to the source and detector, the total probability distributions become harder to evaluate. Usually this is handled by estimating worst cases and setting a decision threshold accordingly. Handbooks with values of the Gaussian or Poisson distribution for values well out on the tail of the curve are readily available. It is not uncommon to find criteria in use such as a chance of one in 1000 for false rejection, and one in 100,000 for false acceptance, with counting times per sample of a fraction of a second. As a specific example, if we have a situation where the mean counting rate when the element is present is 100 counts per second, and 50 counts per second when the element is absent (remember that the tag element is usually a trace element both for economic reasons and to have minimal effect on the host material, and that consequently the count rates will not be high), a counting time of 1.5 seconds and a decision threshold of 102 counts will give expected error rates of one false rejection in over 1000, and one false acceptance in over 20,000 measurements.

Examples of situations where the use of this technique have been considered include the attaching of tags to parts to be sorted later, such as laundry items from a number of different clients which are otherwise indistinguishable, or injection of inert polymers containing tag elements into fish for identification, and the incorporation of the tag elements into a product formulation. An example of the latter is in the manufacture of 'O' ring gaskets from different polymers. The parts are dimensionally alike, and so sorting them could only be based on tag elements added to each polymer batch.

In situations like this, it is necessary that the various tag elements be selected so as not to interfere analytically with each other, or to be interfered with by anything in the sample (it is also assumed, of course, that they are present in low concentration or in some form that does not alter the properties of the material being sorted). Usually, this requires choosing a series of elements which are close together in energy (and hence in the periodic table) so that they can all be efficiently excited with a single source (either an isotope, as in the case of the fish, or an X-ray tube, for the 'O'-rings). For the latter case, oxides of transition metals (Ti, Cr, Fe, etc.) were added in concentrations of a few tenths percent. In many other cases, oxides of the rare earth elements are used as being unlikely to occur by chance, and having K-shell lines that can be excited with an isotope and will be relatively unaffected by absorption in the parts being sorted. It is necessary to choose enough elements to allow sorting of all the possible classes of things to be encountered, of course. If there are a total of N elements available, to be used in combinations of M at a time, then the total number of combinations is N! / (N-M)! M! where the ! indicates factorial. For instance, with 7 possible elements, used in sets of three at a time, it is possible to have 5040 / 6×24 or 35 different combinations.

Alloy Sorting

A very different type of sorting problem arises when many naturally occuring elements are present in material in quite different concentrations, and we wish to sort them based on the intensities. In this case, instead of simply dealing with the statistics of detecting or not detecting the element, we are looking at distingushing various levels of concentration, which are furthermore not generally spaced apart to provide optimal X-ray statistics. The most common application of this technique is to the sorting of metal alloys, either in production or receiving and quality assurance operations, including scrap metal sorting before remelting it. Although the terminology used corresponds to this application, it is important to note that this is far from the only use of the technique. In particular, there have been several attempts to apply this sorting method to forensic identification of automobile manufacturer, type and year from the pattern of pigment and other elements in paint fragments found after accidents. This has met with limited success, due to the enormous variability in the paint formulations used. The matching of spectra from accident fragments and suspect vehicles, which also should properly include statistical matching considerations, has been much more successful.

Sorting among many classes of alloys based on measured intensity values must first of all deal with the practical consideration that the pieces to be analyzed are rarely the ideal flat surfaces needed for classic quantitative analysis. Instead, the surface is likely to be irregular and rough, possibly covered with some oxidation, and the pieces may not all be equally large. Sometimes, in fact, the sample to be identified will be nothing more than some filings or material embedded on a sanding disk carried in from a remote warehouse or stockyard. In these cases, it is impossible to use the intensities directly. Ratios of element to element (for instance, in steels, the ratio of each alloy element to the matrix element iron) or of each element to the total of all analyzed elements (in other words, the normalized intensities) are used instead. These are largely insensitive to sample size, and for moderately high energy elements are even able to cope with variation in surface roughness and oxidation or other coatings, within the variation in intensity we shall expect and allow for repeated analysis of the same type of material.

Most alloy sorting operations are carried out using the normalized intensities directly, as these require no computation and can often be directly compared to intensities measured on standard representatives of each alloy type (Russ, 1973). We shall shortly consider schemes in which psuedo-concentrations are derived from the intensities before searching is done. Let us begin by imagining a three dimensional space, where the three axes represent normalized intensities for the three elements in a simple ternary system (for instance, iron,

285

chromium and nickel in stainless steels). Clearly, we shall be able to extend this later to higher dimensions, but these are difficult to draw or visualize. Figure 25.2 shows this space (the axes could equally well be Cr/Fe , Ni/Fe and another element, say Mo/Fe).

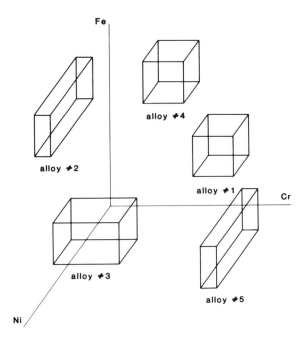

Figure 25.2 Intensity space with regions assigned to various known alloys.

As drawn, several boxes are represented to show ranges of intensity which are to be identified as particular alloys. The measured normalized intensities (or intensity ratios) define a point in this space. If the point lies within one of the boxes, we shall claim that particular alloy has been identified. This is indeed the way that many alloy sorting programs work. Consequently, the size of the box must reflect all of the various sources of variation in allowable intensity, which include the statistics of the measurement and the variation in the concentration of the alloy. The establishment of the limits requires in general the actual analysis of several specimens of each alloy, with known actual composition. Then, from the known composition and the measured intensities, and knowing the composition limits assigned to the alloy type, it is possible to estimate the intensity limits of the box. This estimation makes implicit use of the assumption that intensity is linearly proportional to concentration. Usually, first the intensities measured for a comparatively long time (compared to that which will be later used to measure intensities from unknowns) are obtained, used with the known composition to determine the limits, which are then expanded to take into account the statistical error expected in the intensities when the actual analysis time is used.

There are several real problems with this scheme, having to do with measured normalized intensity values from unknowns which do not fall exactly within an alloy box. First, we would ideally like to have some better way to separately handle the uncertainty due to actual composition variation and counting statistics, so that a longer analysis time might be used when needed to resolve some of the uncertainty. This could be done by having the boxes represent just the variation due to composition range, and each measured point have a finite size reflecting its counting statistics. The problem is then to find out if any part of the "point" overlaps any of the boxes, which requires a very small amount of additional computation.

Next, we recognize that some unknowns may not be within the specified concentration ranges for the alloys, at least as estimated by the boxes, but we would still like to know which alloy is most like the unknown. This can be expressed in more than one way. For instance, is the "nearest" alloy the one which differs from the measured unknown in the fewest elements, or is it the box which is closest in a vector sense, that is which has the smallest total difference in the sum of squares of differences of all elements. The latter seems more reasonable, but is less easy to estimate in this intensity space because the relationship between intensity and concentration is non-linear. This is not simply due to the non-linearity of matrix absorption and enhancement, as were discussed before, but also to the fact that the intensities measured and normalized contain some background and possibly some overlap intensity from other elements (for instance, the Cr K-β is added to the Mn K-α). This off set from zero, and the effect of the normalization, serves to make extrapolation outside the boxes somewhat dangerous.

Another problem with the simple intensity approach is that it becomes difficult to add additional alloy types, since for each one a few representative known specimens must be found and measured. Also, the intensities (normalized or ratios) depend strongly on the particular instrument used for analysis, and cannot be readily shifted from one instrument or set of analysis conditions to another.

Several solutions, of varying complexity and success, have been proposed for this problem. One is to process the measured spectrum just far enough to roughly estimate background (say, by straight line interpolation beneath the peaks), and to use permanently stored overlap factors to correct for elemental line overlaps (Russ,1980b). This is in fact adequate in most cases to determine 'pure' elemental intensities, free of background and overlap, with sufficient accuracy considering the poor counting statistics a vailable in the brief time used for alloy sorting anyway. It permits some extrapolation of intensity ranges from composition ranges, but does not solve the problem of finding the "nearest" alloy, or of adding new alloys to the list.

Using fundamental parameters methods (discussed before) it is possible to compute from first principles the intensities that would be obtained from a sample of any specified composition. This is time consuming, but need be done only once, when a new a lloy class is to be added to the file. If the alloy sorting system has a sufficient computer attached, it can be programmed to carry out this calculation. The question of finding "nearest" alloys, or of data compatibility between instruments remains.

To solve this problem, it is necessary to get out of the arbitrary intensity space so far described, and into concentration space. We have seen several methods for converting intensity to concentration, ranging from simple linear calibration curves to regression mehods to full fundamental parameters calculations. In principle, any could be used. In practice, the more elaborate ones are not, because of the time and computing power needed to carry them out (especially the iterative ones). Instead, either a linear relationship between intensity and concentration, using a stored factor for each element multiplied by the net intensity estimated as before, or a simple intensity correction model is used. The latter requires very little more computation than a linear calibration curve, for each unknown, and includes a built in way to handle background and overlaps based on the regressed factors (the nature of the intensity – correction or Lucas-Tooth method was described in an earlier chapter). It does require some regression beforehand, using measured intensities from known alloys. The normalization of intensities is still practical to correct for sample size and shape.

With either method, it is easy to add more alloys, since only their concentration ranges need be entered into the file. The computation time with a small computer or dedicated microprocessor is well under a second, and sorting for matches goes on as before except that the uncertainty from the counting statistics must be propagated into uncertainty in concentration. Note that the concentration determined after a few seconds of analysis is still far from a quantitative result, but it is not intended to be. The size of the "point" that represents the unknown in our multi-dimensional concentration space may be quite large, as long as it does not overlap more than one alloy type.

In case no alloy type is intersected, and there is no "match", the nearest alloy can now be specified as the one which is closest in a vector sense. The distance from the surface of the ellipsoid defined as the measured point with its statistical errors (rather than its center) to the nearest point on each box should give a reliable estimate of the total difference between the expected result and the various alloys in the file. If one or more elements are to be ignored in this search for the nearest alloy, it is a simple matter to leave them out of the sum of squares of differences (this is akin to finding the distance between the measured unknown point and the boxes in one plane, for our three dimensional example above – eliminating one element simply projects the figure onto a space of one lower dimension).

Systems using these schemes for alloy sorting and classification are routinely used in industry, particularly for scrap metal sorting or incoming alloy certification. Some even can proceed automatically to extend the analysis time when necessary to resolve particular uncertainties, or to perform classic quantitative analysis when needed (using a longer time, of course). Either X-ray tubes, exciting the sample with continuum, or radioactive sources are generally used. The former does a good job of simultaneously analyzing all important elements, which for alloy sorting purposes rarely include minor or trace elements. The latter is comparatively inexpensive, and the difficulty in exciting the lower atomic number elements with practically available isotopes is not too important since the systems used for rapid alloy sorting do not usually measure the light elements anyway (for one thing, they do not usually have an evacuated sample chamber).

Extension of this technique to classification of particles or phases in samples examined in the electron microprobe or electron microscope is direct and straightforward, and was described in an earlier chapter. The principle difference is that the measured intensities are much lower with practical focussed electron beams, and the computation of concentration is not simple, so sorting based on intensities rather than concentrations is more practical.

Matching Stored Spectra

Another type of matching, which is often requested by novice users (either with electron or X-ray excitation), is to find the previously measured spectrum (presumably stored on disk by the computer inside the X-ray analyzer) which "is most like" the current spectrum. Actually, this is rarely used in practice since most users find that their own recognition of spectral patterns is likely to be much faster than the computer searching through disk storage (plus the fact that storage is quite finite with these systems, and you simply cannot save every spectrum you measure). Also, such comparisons are properly only meaningful for spectra collected under similar analysis conditions. For instance, if the electron beam voltage or surface orientation are changed in the SEM or microprobe, or the tube voltage, sample preparation or use of a filter varied in X-ray fluorescence, the spectra from the same sample will appear very different, and will frustrate any search/match method.

Nevertheless, methods have been developed to perform reasonably fast searches through stored spectra. They must take into account the different analysis times, beam or tube current, and sample size that may alter the overall intensity scale, and this may be done for instance by expressing the intensity in each channel or for each element as a fraction of the total intensity in the spectrum. One way to proceed is to compare channel by channel the intensity value in the unknown spectrum and in each comparison spectrum. The total sum of squares of differences, usually weighted by the number of counts as an estimate of the statistical precision, and divided by the number of channels, looks much like the reduced χ^2 value described before in matching spectra with the fitted components, and in fact the same programs can often be used.

This method is quite time consuming, however, and even though the channels with a low number of counts are weighted against in the sum, it is necessary to count them all up in getting there. A much faster way, and one which is also tolerant of spectrum calibration shifts, is just to compare the normalized intensities for a list of elements.

This is very fast, and as in the earlier search match procedure involving alloy ranges, can be adapted to work with semiquantitatively estimated concentrations (in this case the normalized ratios of intensity to calculated pure intensity for the excitation voltage, in the case of electron excitation) as well as intensities, to ease the dependence on excitation conditions. In this, as in the other methods, it is usual to report the nearest several spectra, with some figure of merit describing the similarity. The greatest weakness of this method in particular is that any element missing from the list used for comparison may be present in one spectrum and absent in the other, and yet not effect the match criteria. To use this faster method, it is thus important to use a meaningful list of matching elements, both those required to be present, and those whose absence is important. Finally, it is necessary to reiterate that any search match procedure of this type does little more than present a list of spectra which should be visually compared for gross differences, and then perhaps be subjected to at least semiquantitative conversion to concentration, to compensate for any differences in analysis condition, before deciding that the two materials really are similar, or not.

Chapter 26

Conclusions

The purpose of the descriptions, equations and graphs presented thus far has been to develop for the user of these techniques an intuitive and conceptual image of the physical processes involved in the generation and detection of X-rays. The primary emphasis has been on the use of energy dispersive spectrometers, because they are now rather widespread and newer (and consequently less well documented in the existing literature) than the wavelength dispersive type. As further developments take place, in the directions of handling higher count rates and allowing room temperature operation, for instance, the basic ideas set forth here should still be relevant.

The two principal modes of sample excitation are the use of an electron beam, which can be focussed into a small region to provide selective excitation of local features, and the use of X-rays from an X-ray tube or radioactive isotopic source, which produce excitation over very large volumes (both large lateral areas, and by comparison with electron beam excitation, large depths) in the sample. It deserves reiteration that the analytical purpose in the former case is to measure variations in composition within the sample, and in the latter case, to obtain the average composition of the material, which is presumed homogeneous.

There have been some few attempts to produce analytical instruments which seem to reverse this correlation. For instance, the shallow penetration of electrons gives some advantage in exciting the very light elements, whose X-rays cannot escape from deep in the sample. A few bulk analysis systems have therefore coupled an electron beam source (usually a semi-focussed beam from a conventional electron gun covering perhaps a 1 cm. diameter, but β-emitting isotopes can also be used) with either energy or wavelength detection systems. Such a system requires vacuum both for the reduction of X-ray absorption, and also of course for the electron beam. The difficulty of light element quantitative analysis already described should not be overlooked. Presumeably these instruments met some specific analytical need, but they have not proved popular commercially.

Another example is the use of relatively small diamater (from 0.1 to 1 mm.) X-ray beams, positioned on features of interest in a sample, to produce analysis of local regions. While not capable of the spatial resolution of electron beams, and suffering in terms of intensity since they can be produced only by collimation rather than focussing, these have the advantage over electron beams that detection limits are much improved, due to the absence of continuum background. Suitable combination of a light microscope for feature location, with a positional stage for the sample and an adjustable collimator to control the incident X-ray beam can produce an instrument that meets specific analytical needs. An energy dispersive spectrometer is appropriate because of the low X-ray intensity available in such fine beams.

Further variations in instrument configuration should not surprise us, and with the promised developments in room temperature detectors (which will, by virtue of their small size, permit new arrangements of the hardware) or polarized X-ray excitation (which would require it) we may expect to see such changes. The basic interpretation of the physical processes, the equations describing them, and the goals of the analysis, either qualitative or quantitative, will not change.

Sample Preparation

The subject so far avoided in this discussion has been the need and techniques to prepare specimens for X-ray analysis. It is perfectly true (and manufacturers have somewhat misleadingly emphasized in their advertisements for years) that in some cases, there is no (or at most little) need for any preparation. For instance, the scanning electron microscope examination of a metal fracture, and the subsequent analysis of an inclusion observed near the fracture origin, may require at most cutting the piece to fit within the SEM stage. Some of the techniques described can produce rather good quantitative results for such a sample, even without standards. Similarly, many liquids need only be poured into an appropriate cell, with a thin Mylar bottom, and be placed in an X-ray fluorescence spectrometer to obtain the composition of dissolved elements. Even for irregular bulk specimens (the glaze on a coffee cup seems to be a recurring example), the qualitative determination of the elements present can often be accomplished directly and nondestructively.

Life is not always so simple, however. For electron beam excited microanalysis, the sample must be prepared to meet the following criteria:

1) The feature(s) of interest must be on the exposed surface, produced either by cutting or fracturing for a bulk sample, and within a thin section produced for transmission electron microscopy. For bulk samples, the depth of penetration of the beam must be considered, since penetration through a small feature will result in generation of X-rays from the matrix and subsequent confusion. For thin sections, incorporation of the matrix in the excited region is also a problem. Sometimes physical separation techniques such as extraction replicas of surfaces, or ashing of thin sections, can be used to advantage to remove the surounding matrix.

2) For bulk samples in particular, a conduction path for the electrons to ground must be provided. For materials which are even moderately conducting, there may be no need for any special preparation. For insulators, a thin surface coating of carbon or gold, or other metals, may be applied. This must be thin enough to minimally absorb the

generated X-rays, and yet provide good coverage of what may be a very irregular surface.

3) Specimen mounting, support stages, and indeed all of the surrounding hardware, must be designed to minimize stray radiation which can confuse the analysis.

4) For many materials, particularly organic ones, the preparation process must manage to preserve the localization of the elements in the features of interest. This is far from trivial, and often requires extensive development of cryogenic techniques, or specific chemical precipitation methods. For biological materials, a decade or more of active research in these methods has produced some successful techniques, for some materials, but much remains to be done.

5) Some samples, again particularly the organic ones, may be damaged by the electron beam, with either mass loss from heat damage or migration of selected elements away from the analysis site. Cooling is often an acceptable solution to this problem.

A complete discussion of the methods for sample coating (evaporating or sputtering thin films), or of biological specimen preparation techniques for thin sectioning, for instance, would be quite beyond the scope of this book. These are fields where active developments continue to take place, and the individual researcher is directed to the current literature in his or her own field, and to the proceedings of societies involved with electron microscopy and microanalysis, for the latest information. A recent book: Goldstein et.al. "Scanning Electron Microscopy and X-ray Analysis" (see references) and the proceedings of the Microbeam Analysis Society (published annually as "Microbeam Analysis" by San Francisco Press) or the Electron Microscopy Society of America (proceedings published annually by Claitor's Publishers) are particularly useful, as is the Journal of Microscopy, published for the British Royal Microscopical Society by Blackwell.

For bulk X-ray fluorescence analysis, specimen preparation methods are for the most part reasonably well established and documented. The goals of the preparation process are:

1) To assure homogeneity of the sample on a scale small with respect to the depth analyzed, either by fine grinding and pressing, dilution in a liquid (including glasses which may then be solidified), or other means.

2) To produce a flat, uniform surface for analysis.

3) To prevent the loss of any of the elements of interest.

4) Above all, to be absolutely reproducible, so that whatever biases may be introduced in preparation will apply equally to standards and unknowns, and to today's work and tomorrow's, so that it will be incorporated into the calibration curves (or their mathematical representations).

Excellent textbooks, such as those of Bertin and Muller (cited in the references), summarize many of the techniques extant, and furthermore list many of the specific applications already reported in the literature. These, plus the proceedings of conferences on analytical chemistry and specifically X-ray analysis are recommended. For instance, the proceedings of the Denver X-ray Conference, now more than thirty years old, are published annually in book form ("Advances in X-ray Analysis", Plenum Press).

The reader should of course consult such references before trying to evolve his own technique, but always keeping in mind the physical concepts as a guide to whether or not a specific technique, or adaptation of one, will suit the needs of his or her particular application.

References

C.A. Anderson (1966), M.F. Hasler, in X-ray Optics and Microanalysis (R. Castaing, ed.) Hermann, Paris, p.310

C.A. Anderson (1968), D.B. Wittry, An Evaluation of Absorption Correction Functions for Electron Probe Microanalysis, Brit. J. Appl. Phys. (J. Phys. D: ser 2), vol.1, p.529

W. Bambynek (1972) et.al., X-ray Fluorescence Yields, Auger and Coster Kronig Transition Probabilities, Rev. Mod. Phys. vol.44 no.4, p.716-813

N.C. Barbi (1980), Detectability in Energy Dispersive Microanalysis, in SEM 1980 vol II (O. Johari ed.), SEM Inc, O'Hare IL, p.297-308

A.E. Bence (1968), A.L. Albee, J. Geol., vol 76, p.382

M.J. Berger (1964), S.M. Seltzer, Tables of Energy Losses and Ranges of Electrons and Positrons, Nucl. Ser. Rep. No. 39, NAS-NRC Publ. 1133, Nat. Acad. of Sciences, p.205-208

E.P. Bertin (1970), Principles and Practice of X-ray Spectrometric Analysis, Plenum Press, NY

H.A. Bethe (1930), Ann. Phys. Leipzig, vol.5, p.325

H.E. Bishop (1965), A Monte Carlo Calculation on the Scattering of Electrons in Copper, Proc. Phys. Soc., vol.85, p.855

H.E. Bishop (1974), J. Phys. D.: Appl. Phys., vol.7, p.2009-2020

F. Blum (1973), M.P. Brandt, The Evaluation of the use of a SEM Combined with an EDX Analyzer for Quantitative Analysis, X-ray Spectrometry, vol.2, p.121-124

J.D. Brown (1979a), W.H. Robinson, Quantitative Analysis by $\phi(\rho z)$ Curves, in Microbeam Analysis 1979 (D.E. Newbury, ed.) San Francisco Press, p.238-240

J.D. Brown (1979b), A.P. Von Rosensteil, T. Krisch, Quantitative Carbon Analysis from $\phi(\rho z)$ Curves, in Microbeam Analysis 1979 (op.cit.), p.241-242

E.H.S. Burhop (1955), J. Phys. Radium, vol.16, p.625

J. Byrne (1970), N. Howarth, J. Phys. B, vol.3, p.280

W.J. Campbell (1966), E.F. Spano, T.E. Green, Micro and Trace Analysis by a Combination of Ion-Exchange Resin-Loaded Papers and X-ray Spectrography, Anal. Chem., vol.38, p.987-996

R. Castaing (1951), Thesis, Univ. of Paris, France

R. Castaing (1960), Adv. Electr. Elec. Phys., vol.13, p.353

G. Cliff (1972), G.W. Lorimer, Quantitative Analysis of Thin Metal Foils using EMMA-4: The Ratio Technique, in Proc. 5th European Conf. on E.M., Inst. Phys., London, p.140

A.L. Connelly (1972), W.W. Black, Automatic Location and Area Determination of Photopeaks, Nucl. Instrum. Meth., vol.82, p.141-148

J.A. Cooper (1973), Comparison of Particle and Photon-Excited X-ray Fluorescence Applied to Trace Element Measurements of Environmental Samples, Nucl. Instrum. Meth., vol.106, p.525-538

J.W. Criss (1966), L.S. Birks, in The Electron Microprobe (T.D. McKinley, ed.) John Wiley, New York, p.271

J.W. Criss (1978), L.S. Birks, J.V. Gilfrich, A Versatile X-ray Analysis Program Combining Fundamental Parameters and Empirical Coefficients, Anal. Chem., vol.50, p.33

J.W. Criss (1980), Fundamental Parameters Calculations on a Laboratory Microcomputer, in Adv. in X-ray Anal., vol.23, Plenum Press, NY, p.93-97

L. Curgenven (1971), P. Duncumb, Tube Investments Research Report #303

A.J. Dabrovski (1982), Solid State Room Temperature Energy Dispersive X-ray Detectors, in Adv. in X-ray Anal. vol.25 (J. Russ, et.al., ed.), Plenum Press, p.1-22

P. Duncomb (1966), P.K. Shields, Effect of Critical Excitation Potential on the Absorption Correction, in The Electron Microprobe (op.cit.), Wiley, p.284

P. Duncomb (1968), S.J.B. Reed, The Calculation of Stopping Power and Backscatter Effects in Electron Probe Microanalysis, In Quantitative Electron Probe Microanalysis (K.F.J. Heinrich, ed.), Special Technical Publication 298, National Bureau of Standards (U.S. Dept. of Commerce), p.133-154

P. Duncomb (1969), P.K. Shields-Mason, C. DaCasa, Accuracy of Atomic Number and Absorption Corrections in Electron Probe Microanalysis, Proc. 5th Int'l. Cong. on X-ray Optics and Microanalysis, Springer, Berlin, p.146

C.E. Fiori (1976a), R.L. Myklebust, K.F.J. Heinrich, A Method for Resolving Overlapping Energy Dispersive Peaks in an X-ray Spectrum: Application to the Correction Procedure FRAME-B, Proc. 11th Conf. Microbeam Analysis Society, p.12

C.E. Fiori (1976b), R.L. Myklebust, K.F.J. Heinrich, Prediction of
Continuum Intensity in Energy Dispersive X-ray Microanalysis, Anal.
Chem., vol.48 no.1, p.172-176

C.E. Fiori (1981a), D.E. Newbury, R.L. Myklebust, Artifacts Observed
in Energy Dispersive X-ray Spectrometry in Electron Beam Instruments -
A Cautionary Guide, in Energy Dispersive X-ray Spectrometry (K.F.J.
Heinrich et.al., ed.), Special Technical Publication 604, National
Bureau of Standards, U.S. Dept. of Comm., p.315-340

C.E. Fiori (1981b), R.L. Myklebust, K. Gorlen, Sequential Simplex: A
Procedure for Resolving Spectral Interferences in Energy Dispersive X-
Ray Spectrometry, in Energy Dispersive X-ray Spectrometry (NBS 604
op.cit.), p.238-272

C.E. Fiori (1982), C.R. Swyt, J.R. Ellis, Theoretical Characteristic
to Continuum Ratio in Energy Dispersive Analysis in the Analytical
Electron Microscope, in Microbeam Analysis 1982 (K.F.J. Heinrich,
ed.), San Francisco Press, p.57-71

R.D. Giauque (1979), R.A. Garrett, L.Y. Goda, Determination of Trace
Elements in Light Element Matrices by X-ray Fluorescence Spectrometry
with Incoherent Scattered Radiation as an Internal Standard, Anal.
Chem., vol.51, p.511-515

B.C. Giessen (1968), G.E. Gordon, X-ray Diffraction: New High Speed
Technique Based on X-ray Spectrography, Science, vol.159, p.973-975

M. Green (1961), V.E. Cosslett, The Efficiency of Production of
Characteristic X-radiation in Thick Targets of a Pure Element, Proc.
Phys. Soc. (London), vol.78, p.1206-1214

J.I. Goldstein (1976), G.W. Lorimer, G. Cliff, Quantitative X-ray
Analysis of Thin Films in the Electron Microscope, Proc. 12th Annual
Conf. Microbeam Analysis Society, p.25A-C

J.I. Goldstein (1978a), D.B. Williams, Spurious X-rays Produced in the
STEM, SEM/1977 vol.1, (O. Johari, ed.), SEM Inc, O'Hare IL, p.477-482

J.I. Goldstein (1978b), D.B. Williams, in SEM/1977 (op.cit.), p.651

R. Gunnink (1972), J.B. Niday, Computerized Quantitative Analysis by
Gamma Ray Spectrometry, Lawrence Livermore Laboratory UCRL report
51061, vol.1, U.C. Berkeley, CA

T.A. Hall (1974), P.S. Peters, Quantitative Analysis of thin Sections,
and the Choice of Standards, in Microprobe Analysis as Applied to
Cells and Tissues, (T. Hall, P. Echlin, R. Kaufmann, ed.), Academic
Press, London, p.229-237

T.M. Hare (1982), D. Batchelor, J.C. Russ, Multipoint X-ray Analysis
by use of the Backscattered Electron Signal as a Guide, in Microbeam
Analysis 1982 (op.cit.) p.491-494

K.F.J. Heinrich (1966), in The Electron Microprobe, (T. McKinley et.al., ed.), Wiley, NY p.296

K.F.J. Heinrich (1970), Present State of the Classical Theory of Quantitative Electron Probe Microanalysis, National Bureau of Standards Technical Note 621, U.S. Dept of Commerce

K.F.J. Heinrich (1972), Errors in Theoretical Correction Systems in Quantitative Electron Probe Microanalysis - A Synopsis, Anal. Chem., vol.44, p.350-354

K.F.J. Heinrich (1976), D.E. Newbury, H. Yakowitz (ed.) Use of Monte Carlo Calculations in Electron Probe Microanalysis and Scanning Electron Microscopy, National Bureau of Standards Special Publication 460, U.S. Dept of Commerce

K.F.J. Heinrich (1981), Electron Beam X-ray Microanalysis, Van Nostrand

J.M. Jaklevic (1972), F.S. Goulding, D.A. Landis, High Rate X-ray Fluorescence Analysis by Pulsed Excitation, IEEE Trans. Nucl. Sci., vol.3, p.392-397

A. Janossy (1979), K. Kovacs, I. Toth, Parameters for the Ratio Method by X-ray Microanalysis, Anal. Chem., vol.51, p.491-495

G.G. Johnson (1970), E.W. White, X-ray Emission Wavelength and keV Tables for Nondiffractive Analysis, ASTM Data Series DS46, Philadelphia, PA.

K. Kanaya (1972), S.O. Kayama, J. Phys. D.: Appl. Phys., vol.5, p.43

K.J. Karcich (1981), et.al., Utilization of Alternative Chemistry Definition File Structures for Materials Characterization, SEM/1981 (O. Johari, ed.), SEM Inc, O'Hare IL

T.J. Kneip (1972), G.R. Lawson, Isotope Excited X-ray Fluorescence, Anal. Chem., vol.44, p.57A-68A

H.A. Kramers (1923), On The Theory of X-ray Absorption and of the Continuous X-ray Spectrum, Phil. Mag., vol.46, p.836-871

D.F. Kyser (1976), K. Murata, Application of Monte Carlo Simulation to Electron Microprobe Analysis of Thin Films on Substrates, in Use of Monte Carlo Calculations... (NBS 460 op.cit.), p.129-138

G.R. LaChance (1966), R.J. Traill, A Practical Solution to the Matrix Problem in X-ray Analysis, Canadian Spectroscopy, vol.II, p.43-48

D.A. Landis (1970), F.S. Goulding, J.M. Jaklevic, Performance of a Pulsed Light Feedback Preamplifier for Semiconductor X-ray Spectrometers, UCRL Report 19796, U.C. Berkeley Lawrence Radiation Labs

G. Love (1977), M.G.C. Cox, V.D. Scott, A Simple Monte Carlo Method for Simulating Electron-Solid Interactions and its Application to Electron Probe Microanalysis, J. Phys. D.: Appl. Phys., vol.10, p.7-23

H.J. Lucas-Tooth (1961), B.J. Price, A Mathematical Method for the Investigation of Interelement Effects in X-ray Fluorescent Analysis, Metallurgia, vol.64, p.149

J.J. McCarthy (1981), Least Squares Fit with Digital Filters: A Status Report, in Energy Dispersive X-ray Spectrometry (NBS 604, op.cit.), p.273-296

T. Moeller (1982), A Survey of Peak Extraction Algorithms, Bits & Bytes, Tracor Northern, vol.8, no.1, p.20-25

S. Moll (1977), N. Baumgarten, W. Donnelly, Geometrical Considerations for ZAF Corrections in the SEM, Proc. 13th Annual Conf. Microbeam Analysis Society, p.33

A.J. Morgan (1975), T.W. Davies, D.Z. Erasmus, Analysis of Droplets from Iso-Atomic Solutions as a Means of Calibrating a Transmission Electron Analytical Microscope, J. Microscopy, vol.104, p.271-280

H.G.J. Moseley (1913), Phil. Mag., vol.26, p.1024

R.O. Muller (1972), K. Keil, Spectrochemical Analysis by X-ray Fluorescence, Plenum Press, NY

R.G. Musket (1977), Research & Development, vol.28 no.10, p.26

R.G. Musket (1981), Properties and Applications of Windowless Si(Li) detectors, in Energy Dispersive X-ray Spectrometry (NBS 604 op.cit.), p.97-126

R.L. Myklebust (1976), D.E. Newbury, H. Yakowitz, NBS Monte Carlo Electron Trajectory Calculation Program, in Use of Monte Carlo Calculations... (NBS 460 op.cit.), p.105-128

R.L. Myklebust (1977), C.E. Fiori, K.F.J. Heinrich, FRAME-C: A Compact Procedure for Quantitative Energy Dispersive Electron Probe X-ray Analysis, Proc. 8th Int'l. Conf. on X-ray Optics, p.18

R.L. Myklebust (1981), C.E. Fiori, K.F.J. Heinrich, Spectral Processing Techniques in a Quantitative Energy Dispersive X-ray Microanalysis Procedure (FRAME-C), in Energy Dispersive X-ray Spectroscopy (NBS 604 op.cit.), p.365-380

D.E. Newbury (1976), H. Yakowitz, Studies of the Distribution of Signals in the SEM/EPMA by Monte Carlo Electron Trajectory Calculations, in Use of Monte Carlo Calculations... (NBS 460 op.cit.), p.15-44

L. Parobek (1978), J.D. Brown, X-ray Spectrometry, vol.7, p.26

J. Philibert (1963), Proc. 3rd Int'l. Conf. on X-ray Optics and Microanalysis (H. Pattee, et.al., ed.), Academic Press, NY, p.379

J. Philibert (1968), R. Tixier, Electron Penetration and the Atomic Number Correction in Electron Probe Microanalysis, Brit. J. Appl. Phys. (J. Phys. D.), vol.1, p.685

C.J. Powell (1976), Evaluation of Formulas for Inner Shell Ionization Cross-Sections, in Use of Monte Carlo Calculations... (NBS 460 op.cit.), p.97-104

T.S. Rao-Sahib (1972), D.B. Wittry, The X-Ray Continuum from Thick Targets, Proc. 6th Intern. Conf. on X-ray Optics and Microanalysis, p.131-137

S.D. Rasberry (1974), K.F.J. Heinrich, Calibration for Interelement Effects in X-ray Fluorescence Analysis, Anal. Chem., vol.46, p.81-89

S.J.B. Reed (1965), Characteristic Fluorescence Corrections in Electron Probe Microanalysis, Brit. J. Appl. Phys., vol.16, p.913-926

S.J.B. Reed (1966), Spatial Resolution in Electron Probe Microanalysis, 4th Int'l. Conf. on X-ray Optics and Microanalysis, (R. Castaing, ed. op.cit.), p.339

S.J.B. Reed (1971), Rev. Phys. Techn., vol.2, p.92

S.J.B. Reed (1975a), Electron Microprobe Analysis, Cambridge University Press

S.J.B. Reed (1975b), The Shape of the Continuous X-ray Spectrum and Background Corrections for Energy Dispersive Electron Microprobe Analysis, X-ray Spectrometry, vol.4, p.14-17

W. Reuter (1972), Proc. 6th Int'l. Conf. on X-ray Optics and Microanalysis (G. Shinoda, et.al., ed.), Univ. Tokyo Press, p.121

J.R. Rhodes (1966), Radioisotope X-ray Spectrometry (A Review), The Analyst, vol.91, p.688-699

W.H. Robinson (1982), J.D. Brown, Correction for Electron Incidence Angle in Quantitative Analysis, Microbeam Analysis 1982, op.cit., p.159-160

G.W. Roomans (1977), H.L.M. Vangaal, Organometallic Compounds as Standards for Microprobe Analysis of Epoxy Resin Embedded Tissue, J. Microscopy, vol.109 pt.2, p.235-240

J.C. Russ (1973), A.O. Sandborg, M.W. Barnhart, Energy Dispersive Fluorescence Analysis: Use of Multi-Element Detection for Rapid On-Line Classification and Sorting, Proc. XVII Colloquium Spectroscopicum Internationale, Florence, Italy

J.C. Russ (1974a), The Direct Element Ratio Model for Quantitative Analysis of Thin Sections, in Microprobe Analysis of Cells and Tissues, op.cit., p.269-276

J.C. Russ (1974b), Quantitative Microanalysis with Minimum Pure Element Standards, Proc. 10th Annual Conf. Microbeam Analysis Society, Ottawa, Canada

J.C. Russ (1975), A Simple Correction for Backscattering from Inclined Surfaces, Proc. 11th Annual Conf. Microbeam Analysis Society, Las Vegas

J.C. Russ (1976), Errors Introduced in Eliminating or Extrapolating Standards, Proc. 12th Annual Conf. Microbeam Analysis Society, p.19

J.C. Russ (1977a), Measuring Detector Entrance Windows, Proc. 13th Annual Conf. Microbeam Analysis Society, p.105

J.C. Russ (1977b), Use of the Electron Backscatter Factor to Normalize X-ray Microanalysis, Proc. 13th Annual Conf. Microbeam Analysis Society, p.34

J.C. Russ (1978), A Fast, Self Contained No-Standards Quantitative Program for EDS, Proc. 14th Annual Conf. Microbeam Analysis Society, p.46

J.C. Russ (1979), New Methods to Obtain and Present SEM X-ray Line Scans, Microbeam Analysis 1979 (op.cit.), p.292-304

J.C. Russ (1980a), T.M. Hare, Quantitative X-ray Microanalysis on Surfaces with Unknown Orientation, Canadian Jnl. Spectroscopy, vol.25 no.4, p.98-105

J.C. Russ (1980b), T.M. Hare, F.U. Luehrs, A Novel and Efficient Alloy Sorting Algorithm, Proc. Denver X-ray Conf., Denver Univ, p.130-131

J.C. Russ (1981a), A.O. Sandborg, Use of Windowless Detectors for Energy Dispersive Light Element X-ray Analysis, in Energy Dispersive X-ray Spectrometry (NBS 604 op.cit.), p.71-96

J.C. Russ (1981b), Multiple Least Squares Fitting for Spectrum Deconvolution, in Energy Dispersive X-ray Spectrometry (NBS 604 op.cit.), p.297-314

J.C. Russ (1981c), T.M. Hare, Characterization of Heterogeneous Polycrystalline Materials, Microbeam Analysis 1981 (R.H. Geiss, ed.), San Francisco Press, p.186-189

R.W. Ryon (1982), J.D. Zahrt, P. Wobrauschek, et.al., The Use of Polarized X-rays for Improved Detection Limits in Energy Dispersive X-ray Spectrometry, in Advances in X-ray Analysis vol.25 (op.cit.), p.63-74

A. Savitsky (1964), M.J.E. Golay, Smoothing and Differentiation of Data by Simplified Least Squares Procedures, Anal. Chem., vol.36, p.1627-1639

F.H. Schamber (1977), A Modification of the Linear Least-Squares Fitting Method which Provides Continuum Suppression, in X-Ray Fluorescence Analysis of Environmental Samples (T.G. Dzubay, ed.), Ann Arbor Science Publ., p.241-257

F.H. Schamber (1981), Curve Fitting Techniques and their Application to the Analysis of Energy Dispersive Spectra, in Energy Dispersive X-ray Spectrometry (NBS 604 op.cit.), p.193-232

J. Scheer (1977), L. Voet, U. Watjen, et.al., Comparison of Sensitivities in Trace Element Analysis Obtained by X-ray Excited X-ray Fluorescence and Proton Induced X-ray Emission, Nucl. Instrum. Meth., vol.142, p.333-338

L.A. Schwalbe (1981), H.J. Trussel, Maximum A Posteriori Technique Applied to Energy Dispersive X-ray Fluorescence Spectra, X-ray Spectrometry, vol.10 no.4, p.187-192

R.B. Shen (1977), A Simplified Fundamental Parameters Method for Quantitative Energy Dispersive X-ray Fluorescence Analysis, X-ray Spectrometry, vol.6, p.56

J. Sherman (1955), The Theoretical Derivation of Fluorescent X-ray Intensities from Mixtures, Spectrochim. Acta, vol.7, p.283

J. Sherman (1959), Simplification of a Formula in the Correlation of Fluorescent X-ray Intensities from Mixtures, Spectrochim. Acta, vol.15, p.466

R. Shimizu (1977), Some Recent Developments in Microbeam Analysis in Japan-1) Application of Monte Carlo Calculation to Fundamentals of Microbeam Analysis, Technology Reports, Osaka University, vol.27 no.1343, p.69-90

G. Shinoda (1968), K. Murata, R. Shimizu, Scattering of Electrons in Metallic Targets, in Quantitative Electron Probe Microanalysis (NBS 298 op.cit.), p.155-188

T. Shiraiwa (1966), N. Fujino, Theoretical Calculation of Fluorescent X-ray Intensities in X-ray Spectrochemical Analysis, Japanese Jnl. of Applied Physics, vol.5, p.886

J.A. Small (1979), D.E. Newbury, R.L. Myklebust, Analysis of Particles and Rough Surfaces by FRAME-P, a ZAF Method Incorporating Peak to Background Measurements, in Microbeam Analysis 1979, op.cit., p.243-253

D.G.W. Smith (1976), C.M. Gold, A Scheme for Fully Quantitative Energy Dispersive Microprobe Analysis, in Adv. in X-ray Analysis vol.19, Plenum Press, p.191-201

L.F. Solosky (1972), D.R. Beaman, A Simple Method for Determining the Acceleration Potential in Electron Probe and Scanning Electron Microscopes, Rev. Sci. Instrum., vol.43 no.8, p.1100-1102

A.R. Spurr (1974), Macrocyclic Polyether Complexes with Alkali Elements in Epoxy Resin as Standards for Microprobe Analysis of Biological Tissues, in Microprobe Analysis as Applied to Cells and Tissues, op.cit., p.213-227

P.J. Statham (1976), Escape Peaks and Internal Fluorescence for a Si(Li) detector and General Geometry, J. Phys. E.: Sci. Instrum., vol.9

P.J. Statham (1978), J.B. Pawley, A New Method for Particle X-ray Microanalysis Based on Peak-to-Background measurements, SEM/1978 (O. Johari, ed.), SEM Inc, O'Hare IL, p.469-478

P.J. Statham (1981), X-ray Microanalysis with Si(Li) Detectors, J. Microscopy, vol.123, p.1-23

P.J. Statham (1982), Electronic Techniques for Pulse Processing with Solid State X-ray Detectors, in Energy Dispersive X-ray Spectrometry (NBS 604 op.cit.), p.141-164

D.A. Stephenson (1971), Theoretical Analysis of Quantitative X-ray Emission Data: Glasses Rocks and Metals, Anal. Chem., vol.43, p.934

P.M. Van Dyck (1980), R.E. Grieken, Absorption Correction via Scattered Radiation in Energy Dispersive X-ray Fluorescence Analysis for Samples of Variable Composition and Thickness, Anal. Chem., vol.80, p.1859-1864

R.A. Vane (1980), W.D. Stewart, The Effective Use of Filters with Direct Excitation of Energy Dispersive X-ray Fluorescence, in Advances in X-ray Analysis vol.23 (J.R. Rhodes, et.al. ed.), Plenum Press, p.231-239

J.N. Van Niekerk (1960), J.F. DeWet, Trace Analysis by X-ray Fluorescence using Ion Exchange Resins, Nature, vol.186, p.380-381

G. Wentzel (1927), Z. Phys., vol.43, p.524

R.R. Wilson (1941), Phys, Rev., vol.60, p.749

R. Woldseth (1973), Energy Dispersive X-ray Spectrometry, Kevex Corp., Burlingame CA

H. Yakowitz (1973), R.L. Myklebust, K.F.J. Heinrich, FRAME: An On-Line Correction Procedure for Quantitative Electron Probe Microanalysis, National Bureau of Standards Technical Note 796, U.S. Dept. of Commerce

Author's Note: This list of references is intended to document the specific equations and techniques cited in this presentation of the development and current state of the art for Energy Dispersive X-ray Analysis, but not to form a comprehensive bibliography for the field. Principal emphasis has here been placed on either seminal or review articles. Excellent bibliographies are available in several places, including:

A bibliography of publications relating principally to energy dispersive X-ray analysis with electron beams, in National Bureau of Standards publication 604 (cited above).

Extensive references to the fundamental mathematical relationships involved in quantitative correction procedures for electron microprobe analysis in Heinrich's 1981 book (cited above).

References to the various mathematical models used in X-ray fluorescence corrections, as well as indices to a broad variety of sample preparation techniques and specific analytical applications, in the books by Muller and Bertin (cited above).

The interested reader is also directed in particular to the annual proceedings of the Microbeam Analysis Society, published by San Francisco Press, for articles relating to electron excitation, and to the series Advances in X-ray Analysis (the proceedings of the annual Denver X-ray Conference), published by Plenum Press. Most of the important literature in this field either appears or is cited there.

Index

LaChance, G.R., 223
LaChance–Traill model, 223–225
Layered structures,
    analysis, 244–245
Lenard coefficient, 75
Library spectra, 140–141
Line intensities, 3–4
Linear least squares fitting method, 138–141
Love, G., 60
Lucas-Tooth, H.J., 225

McCarthy, J.J., 138, 141
Mass absorption coefficients, 70–73
    errors, 122
    matrix 256–257
Matrix absorption,
    scattered intensity estimation, 258
Mean ionization potential, 42–43
Microprobes,
    stability, 101
Minerals,
    standards, 169–170
Moeller, T., 41
Molybdenum,
    target element, 210
Monte Carlo modelling, 57–68
Morgan, A.J., 170
Moseley, H.G.J., 4
Muller, R.O., 223, 241, 253–254
Multi-channel analyzers, 34–37
Musket, R.G., 279, 281
Myklebust, R.L., 66–67, 134, 149, 150

Newbury, D.E., 62–65

Overlap factors, 148–151
Overvoltage, 108–110, 127

Parobek, L., 48
Particle size,
    sample heterogeneity, 238–241
Particles,
    analysis, 63–64
Peak to background modelling, 161–164
Peak to background ratio,
    thin sections, 165–168
Peaks,
    deconvolution, 138–141
    escape, 29–30
    identification, 41
    overlaps, 39–41, 148–151
    profiles, 38, 141–143
    shapes, 38
    spectra, 4–5
Phase boundaries,
    secondary fluorescence effects, 82
Phase compositions,
    analysis, 193–200
Philibert, J., 75
Photon excitation, 8–9, 209–211
Photons, 1
PIXE, 276–279
Polarization,
    X-rays, 273–275
Powders, 239–241
Powell, C.J., 107
Primary absorption effect, 214
Process monitoring,
    X-ray analysis, 282–283

Proportional counters,
    gas, 13–16
Proton induced X-ray emission, 276–279
Pulses,
    amplification, 21–23
    counting, 34–37
    measurement, 32–33
    pile–up, 21–23

Qualitative analysis, 37–41, 234–237
Quality control,
    X-ray analysis, 282–283
Quantum numbers, 3

Radioisotopes,
    sources, 266–268
Rao-Sahib, T.S., 155
Rasberry, S.D., 226
Ratemeters,
    X-ray maps, 188–191
Rayleigh scatter, see Coherent scatter
Reed, S.J.B., 44, 45, 53, 86, 93, 122, 153, 157
Reference spectra, 140–141
Resolution,
    energy dispersive spectrometry, 27–28
    thin sections, 172
Reuter, W., 50
Rhodes, J.R., 266
Robinson, W.H., 50
Roomans, G.W., 169
Russ, J.C., 53, 56, 112, 115, 124, 170, 189, 204,
    207, 286
Ryon, R.W., 273, 275

Sample effects,
    X-ray fluorescence, 238–247
Sample preparation, 293–294
    grinding/compaction, 238–241
Samples,
    heterogenous, 193–200
Savitsky, A., 131, 136
Scanning electron microscope, 17
    stability, 101
Scanning transmission electron microsopy, 25,
    25–26, 165–174, 208
Scattering, 248–257
    large angle, 49–54
    modelling, 57–58
    X-rays, 70–73
    see also Coherent scatter, Incoherent scatter
Schamber, F.H., 137
Scheer, J., 277
Secondary absorption, 213
Secondary fluorescence, 81–90, 214–217
Secondary targets, 260–266
Seltzer, S.M., 42
Sheet steel,
    surface analysis, 202
Sherman, J., 217
Shields, P.K., 75
Shimizu, R., 59
Shinoda, G., 47
Shiraiwa, T., 217
Signals,
    stray, 159–161
Small, J.A., 163
Smith, D.G.W., 149
Smoothing,
    spectrum data, 130–131

307